Ethical Artificial Intelligence in Power Electronics

This book focuses on the techniques of artificial intelligence that are mainly used in the power electronics field for the optimization of lost vehicle power. With the intention of optimizing the powerful energy of the vehicles and producing reliable energy, the most efficient methods, algorithms, and strategies of ethical artificial intelligence (AI) are being applied. By employing machine learning methods, the optimization of power energy in vehicles can be quickly recovered and managed efficiently. In today's bustling world, power energy is indispensable for progress, yet in congested Vehicular Ad-hoc Networks (VANETs), vehicles often face power depletion and decreased efficiency. This book explores these challenges, encompassing not only power but also other critical power electronics within vehicles. We aim to introduce innovative approaches, leveraging ethical AI methods, to optimize energy performance in the face of these difficulties. Through this exploration, we seek to provide practical insights into navigating congested VANET environments while upholding ethical principles in technological advancements. Our book will discuss the current power energy concerns faced by vehicles and also contribute a novel strategy to overcome those concerns. The employment of ethical AI in vehicular power energy will undoubtedly improve the effectiveness and production of vehicles.

Ethical Artificial Intelligence in Power Electronics

Edited by
Tarandeep Kaur Bhatia, Salma El Hajjami, Keshav Kaushik,
Gayo Diallo, Mariya Ouaissa, and Inam Ullah Khan

CRC Press
Taylor & Francis Group
Boca Raton London New York

CRC Press is an imprint of the
Taylor & Francis Group, an **informa** business

Designed cover image: Shutterstock Images

First edition published 2025
by CRC Press
2385 NW Executive Center Drive, Suite 320, Boca Raton FL 33431

and by CRC Press
4 Park Square, Milton Park, Abingdon, Oxon, OX14 4RN

CRC Press is an imprint of Taylor & Francis Group, LLC

ISBN: 978-1-032-63115-8 (hbk)
ISBN: 978-1-032-64831-6 (pbk)
ISBN: 978-1-032-64832-3 (ebk)

DOI: 10.1201/9781032648323

Typeset in Minion
by Newgen Publishing UK

Contents

Preface

In an era marked by the omnipresence of technology, our daily lives are intricately intertwined with the advancements that artificial intelligence (AI) brings. As we navigate the complex landscape of technological progress, one crucial aspect that demands our attention is the ethical integration of AI into specific domains, particularly within the realm of power electronics.

This book, "Ethical Artificial Intelligence in Power Electronics," serves as a comprehensive exploration into the techniques of AI applied to power electronics, with a primary focus on optimizing the energy efficiency of vehicles.

Acknowledging power energy as the cornerstone of modern life, our work delves into the challenges faced by vehicles, particularly in the context of vehicular ad-hoc networks (VANETs). In such environments, vehicles often experience power losses and diminishing efficiency, necessitating innovative solutions to address these issues.

This book endeavors to unravel the potential of ethical AI as a transformative force in optimizing vehicle power energy, employing cutting-edge methods, algorithms, and strategies. Through the lens of machine learning, we embark on a journey to swiftly recover and efficiently manage the optimization of power energy in vehicles. This practical guide aims to provide a comprehensive introduction to the unique challenges posed by power energy constraints in the congested conditions of VANETs.

By offering insightful approaches and innovative ethical AI methods, we aim to not only highlight the existing concerns surrounding power energy in vehicles but also present novel strategies to overcome these challenges.

Our book not only identifies the pressing issues related to power energy in vehicles but also contributes a fresh perspective on addressing these concerns through the application of ethical AI strategies. We firmly believe that the incorporation of ethical considerations in the realm of vehicular power energy will undoubtedly enhance the overall effectiveness and productivity of vehicles, paving the way for a more sustainable and optimized future.

This book provides detailed information about ethical AI in power electronics. However, ethical AI/machine learning for IoT, smart cities & VANETs are briefly discussed. Each and every chapter presents technological solutions and opportunities regarding ethical AI in power electronics. This book is a detailed guide for researchers on ethical AI in power systems, smart girds, blockchain, and IoT-based electrical appliances. It is organized as follows:

- **Chapter 1** sheds light on the integration of ethical machine learning with IoT. Recognizing the pivotal role of AI in societal advancement, this chapter emphasizes data-centric decision-making as a catalyst for improving our daily lives.
- **Chapter 2** presents the applications of ethical machine learning in shaping industries, with a particular focus on its transformative impact on healthcare and transportation. The resulting technological integration is poised to usher in a new era of smart cities.
- **Chapter 3** shifts its focus to the merger of IoT with ethical machine learning, unraveling basic concepts, addressing challenges, and presenting real-time applications. This chapter serves as a guide, providing a nuanced understanding of the synergistic potential inherent in the integration of IoT and ethical machine learning, particularly in shaping the future landscape of communication systems.
- **Chapter 4** provides a full overview of "Machine Learning in Wireless Communication Networks", including AI techniques that are applied to develop a smartphone application for various advancements in the upcoming years.
- **Chapter 5** offers a detailed examination of routing protocols, shedding light on the intricate mechanisms that govern data transfer and communication protocols. Simultaneously, our focus extends to the ethical dimensions of machine intelligence within the Internet of Things (IoT) domain, emphasizing the importance of ethical considerations in the development and deployment of intelligent systems.
- **Chapter 6** explains more about healthcare and the ongoing battle against the global pandemic, with a spotlight on AI and blockchain solutions for managing medical data, particularly in the context of Covid-19.
- **Chapter 7** offers a deep exploration of Industry 4.0/5.0, where we navigate the transformative waves of the industrial sector fueled by technological advancements.
- **Chapter 8** discusses the potential of AI-enabled risk assessment methodologies, particularly in the context of construction projects. By leveraging advanced AI algorithms, this chapter seeks to unravel how risk assessment processes can be enhanced and optimized, ensuring a more proactive and responsive approach to potential challenges in construction endeavors.
- **Chapter 9** extends its reach to the IoT, exploring how AI can be harnessed to assess and mitigate risks in the intricate web of interconnected devices and systems.
- **Chapter 10** briefly explains IoT-based electrical appliances. The application of AI in smart grids is also a focal point, where risk assessment becomes pivotal in maintaining the reliability and resilience of energy distribution networks.

Editor Biographies

Tarandeep Kaur Bhatia received her Ph.D. degree in Computer Science Engineering from Deakin University, Australia under Higher Degree by Research Scholarship. Currently, she is working as an Assistant Professor in University of Petroleum & Energy Studies (UPES) Bidholi, Dehradun, India. She is a Gold Medalist in her M.Tech, Computer Science & Engineering batch 2015–2017. She published many papers in reputed journals, presented several papers in many National/International conferences, and also wrote many book chapters. Dr. T.K Bhatia filed 20 Patents. She triumphantly completed and conducted many Certified Faculty Development Programmes (FDP). She successfully completed many certified trainings from the Queensland University of Technology, Australia, University of Leeds (England), and Murdoch University, Australia. Dr. Tarandeep Kaur is a "Member of reviewer committee" for the *IEEE Communications Standards Magazine* Journal, *IEEE Access, IEEE Transactions on Vehicular Technology, Journal of Supercomputing, IET Communications, Physical Communications*, and many more. She is also serving as an "Editorial and Review Board Member" for LC International Journal of STEM (LC-JSTEM) and Panel Member of the Technical Program Committee (TPC) of the COMS2-2021 conference. Her areas of interest are Vehicular Ad-hoc NETworks (VANETs), Data Congestion, Networking, Software Engineering, Artificial Intelligence. Recently working with various simulators and tools such as NS-2, MATLAB®, Omnet++, SUMO, VEINS for her research work.

Salma El Hajjami (Eng., Ph.D.) is an Assistant Professor and Researcher at Faculty of Science, Ibn Zohr University, Agadir, Morocco since 2021. She is a Ph.D., graduated in 2021 in Computer Science, at the Laboratory of Artificial Intelligence, Data Science and Emerging Systems from ENSA, Sidi Mohammed Ben Abdellah University, Fez, Morocco. She is a Computer Science Engineer, graduated in 2015 from National School of Applied Sciences Fez, Morocco. She has previous expertise acting in Ministry of interior Morocco as Research and Development Engineer from 2017 to 2021. She is member of International Association of Engineers (IAENG) and International Association of Online Engineering. Dr. Salma has made contributions in the fields of Social Big Data, Semantics Analytics, Anomaly Detection, and Imbalanced Big Data published at international conferences and journals. Her main research topics are Machine Learning, Deep Learning, Imbalanced Big Data, Data Science and Blockchain. She has served and continues to serve on technical program and organizer committees of several conferences also as a reviewer of numerous international journals.

Keshav Kaushik is an experienced educator with over eight years of teaching and research experience in Cybersecurity, Digital Forensics, and the Internet of Things. He is working as an Assistant Professor (Senior Scale) in the systemic cluster under the School of Computer Science at the University of Petroleum and Energy Studies, Dehradun, India. He has published 80+ research papers in International Journals and has presented at reputed International Conferences. He is a Certified Ethical Hacker (CEH) v11, CQI and IRCA Certified ISO/IEC 27001:2013 Lead Auditor, Quick Heal Academy Certified Cyber Security Professional (QCSP), and IBM Cybersecurity Analyst. He acted as a keynote speaker and delivered 50+ professional talks on various national and international platforms. He has edited over ten books with reputed international publishers like Springer, Taylor, and Francis, IGI Global, Bentham Science, etc. He has chaired various special sessions at international conferences and also served as a reviewer in peer-reviewed journals and conferences.

Gayo Diallo is a full professor in computer science at Bordeaux University (France) since 2022 and is deputy director of the AHeaD research group of the Bordeaux Population Health (BPH) research center (BPH) Inserm-1219. Previously, he was the group leader of the Computer Sciences Applied to Health research group (ERIAS) of BPH. He joined the University of Bordeaux, Bordeaux School of Public Health (ISPED), in 2009 after being a research assistant at City University of London (UK) and PostDoc researcher at the Laboratory of Applied Computer Sciences (LISI/ENSMA) Futuroscope Poitiers (France). He graduated from University of Grenoble Joseph Fourier in 2006 and holds a degree in Business Creation obtained in 2003. His research interests include AI based approaches for healthcare data and knowledge management and ICT for Societal Development (ICT4D). He coordinated several international and regional R&D projects and has been involved or coordinated various EU or other international funded projects. Gayo was awarded a IDEX Talent mobility grant, and he has been Visiting Professor at University of Minnesota/the Institute of Health Informatics (US) during the period January-August 2022 and held for two years a part-time Inria delegation position. He is the winner of the practical application prize of the 2015 edition of the Orange Data for Development Big Data Challenge (D4D) and he authored or co-authored more than 100 peer-reviewed papers on symbolic AI and digital health.

Mariya Ouaissa is currently a Professor in Cybersecurity and Networks at Cadi Ayyad University and practitioner with industry and academic experience. She is a Ph.D. graduated in 2019 in Computer Science and Networks, at the Laboratory of Modelisation of Mathematics and Computer Science from ENSAM-Moulay Ismail University, Meknes, Morocco. She is a Networks and Telecoms Engineer, graduated in 2013 from National School of Applied Sciences Khouribga, Morocco. She is a Co-Founder and IT Consultant at IT Support and Consulting Center. She was working for

School of Technology of Meknes Morocco as a Visiting Professor from 2013 to 2021. She is member of the International Association of Engineers and International Association of Online Engineering, and since 2021, she is an "ACM Professional Member". She is Expert Reviewer with Academic Exchange Information Centre (AEIC) and Brand Ambassador with Bentham Science. She has served and continues to serve on technical program and organizer committees of several conferences and events and has organized many Symposiums/Workshops/Conferences as a General Chair also as a reviewer of numerous international journals. Dr. Ouaissa has made contributions in the fields of information security and privacy, Internet of Things security, and wireless and constrained networks security. Her main research topics are IoT, M2M, D2D, WSN, Cellular Networks, and Vehicular Networks. She has published over 50 papers (book chapters, international journals, and conferences/workshops), 13 edited books, and 8 special issue as guest editor.

Inam Ullah Khan is the Founder of AI-EYS. Recently, he is working as Global Mentor/ Guest Lecturer at Impact Xcelerator, IE School of Science and Technology, Madrid, Spain and SZABIST, Islamabad, Pakistan. Previously, he was working as Visiting researcher at King's College London, United Kingdom. He was faculty member at different universities in Pakistan which include Center for Emerging Sciences Engineering & Technology (CESET), Islamabad, Abdul Wali Khan University, Garden Campus, Timergara Campus, University of Swat & Shaheed Zulfikar Ali Bhutto Institute of Science and Technology (SZABIST), Islamabad Campus. He completed his Ph.D. in Electronics Engineering from Department of Electronic Engineering, Isra University, Islamabad Campus, School of Engineering & Applied Sciences (SEAS). Also, he did his M.S. degree in Electronic Engineering at Department of Electronic Engineering, Isra University, Islamabad Campus, School of Engineering & Applied Sciences (SEAS). He had done undergraduate degree in Bachelor of Computer Science from Abdul Wali Khan University Mardan, Pakistan. Apart from that his Master's thesis is published as a book on topic "Route Optimization with Ant Colony Optimization (ACO)" in Germany which is avaliable on Amazon. He authored/coauthored more than 60 research articles in reputable journals, conferences and book chapters. More interestingly he recently introduced a novel routing protocol E-ANTHOCNET in the area of Drones/ Unmanned Aerial Vehicles. His research interest includes Network System Security, Intrusion Detection, Intrusion Prevention, cryptography, Optimization techniques, WSN, IoT, Mobile Ad Hoc Networks (MANETS), Flying Ad Hoc Networks, and Machine Learning, AI, Unmanned aerial vehicles. He also served in many Interational conferences as Technical program chair, session chair and Technical program committee member. In addition, he served as editor in around 11 books on various topics. Also, he is Guest Editor, Review Editor, Associate Editor, Editorial Board member at reputed international journals. Last year he was General Chair at International Conference on Trends and Innovations in Smart Technologies (ICTIST'22), virtually from London, United Kingdom. More interestingly, he was invited as technology expert many times on Pakistan National Television.

Contributors

Ahthasham
Faculty of Computing
Department of Software Engineering
Capital University of Science Technology
Islamabad, Pakistan

Zaigham Abbas
Department of Electrical & Computer Engineering
International Islamic University
Islamabad, Pakistan

Qasem Abu Al-Haija
Department of Cybersecurity
Faculty of Computer & Information Technology
Jordan University of Science and Technology
Irbid, Jordan

Usama Hussain Ahmad
Department of Electrical Engineering
Qurtuba University of Science & Information Technology
DI Khan, Pakistan

Basheer Ahmed
Department of Electrical & Computer Engineering
International Islamic University
Islamabad, Pakistan

Abdullah Akbar
Department of Electrical & Computer Engineering
International Islamic University
Islamabad, Pakistan

Ahmed Ashfaq
Faculty of Applied Computer Science
Deggendorf Institute of Technology
Deggendorf, Germany

Srinath Doss
Faculty of Engineering & Technology
Botho University
Gaborone, Botswana

Syeda saba Hashmi
University of Karachi
Karachi, Pakistan

Taminul Islam
Daffodil International University
Birulia, Bangladesh

Vijayashree Jayaraman
Vellore Institute of Technology
Vellore, Tamil Nadu, India

Jayashree Jayaraman
Vellore Institute of Technology
Vellore, Tamil Nadu, India

Christian Kaunert
Professor of International Security
Dublin City University
Dublin, Ireland

Maysa Khalil
Department of Data Science
King Hussein School of Computing Sciences
Prince Sumaya University for Technology
Amman, Jordan

Yasir Ali Khan
NUTECH Skills Development Department
National University of Technology
Islamabad, Pakistan

Khalid Mahmood
Faculty of ICT
Department of Computer Science
Baluchistan University of Information Technology Engineering and Management Science
Quetta, Pakistan

Zaid Muhammad
Abdul Wali Khan
University Pabbi Campus
Pabbi, Pakistan

Muhammad Muneer
Department of Robotics and Artificial Intelligence
National University of Science and Technology (NUST)
Islamabad, Pakistan

Mehak Mushtaq Malik
COMSATS University
Islamabad, Pakistan

Mariyam Ouaissa
Laboratory of Information Technologies
Chouaib Doukkali University
El Jadida, Morocco

Nishitha Pujala
Department of Computer Science & Engineering
Stanley College of Engineering & Technology for Women
Hyderabad, India

Zakria Qadir
School of Computing Engineering and Mathematics
Western Sydney University
Penrith, NSW, Australia

Rishi Raghu
Northeastern University
Boston, Massachusetts, USA

Muhammad Tausif Afzal Rana
School of Engineering
Faculty of Science and Technology
Charles Darwin University
Sydney, NSW, Australia

Kishor Kumar Reddy C
Department of Computer Science & Engineering
Stanley College of Engineering & Technology for Women
Hyderabad, India

Faisal Rehman
Department of Statistics and Data Science
University of Mianwali
Mianwali, Pakistan
and
Department of Robotics and Artificial Intelligence
National University of Science and Technology (NUST)
Islamabad, Pakistan

Muhammad Hamza, Sajjad
Department of Statistics and Data Science
University of Mianwali
Mianwali Pakistan

Niusha Shafiabady
Faculty of Science and Technology
Charles Darwin University
Sydney, NSW, Australia

Wasswa Shafik
School of Digital Science
Universiti Brunei Darussalam
Bandar Seri Begawan, Brunei
and
Dig Connectivity Research Laboratory (DCRLab)
Kampala, Uganda

Abdul Siddique
Faculty of ICT
Department of Computer Science
Baluchistan University of Information Technology Engineering and Management Science
Quetta, Pakistan

Bhupinder Singh
Sharda School of Law
Sharda University
Greater Noida, India

Fareed Ud Din
School of Science and Technology
The University of New England
Armidale, NSW, Australia

Aziz Ullah
Faculty of ICT
Department of Computer Science
Baluchistan University of Information Technology Engineering and Management Science
Quetta, Pakistan

Ubaid Ullah
University of Wah
Wah, Cantt, Pakistan

Shahab Ur Rehman
Department of Electrical Engineering
Qurtuba University of Science & Information technology
DI Khan, Pakistan

Muhammad Usama
University of Wah
Wah Cantt, Pakistan

Sai Surya Varshith Nukala
Boston University
Boston, Massachusetts, USA

CHAPTER 1

Ethical Machine Learning for Future Applications

Vijayashree Jayaraman, Rishi Raghu,
Jayashree Jayaraman, and Sai Surya Varshith Nukala

1.1 INTRODUCTION

Machine Learning (ML) consists of computer algorithms that turn data into intelligent action. The development of statistical techniques and processing ability to handle exponential data collection and manipulation are driving growth for smart cities. ML gains knowledge from experience and becomes more intelligent. ML learning and human learning consist of data intake, abstraction, and generalization, but the process is mostly subconscious in humans.

As a result, ML is best thought of as a subset of artificial intelligence (AI), where algorithms guided by sophisticated neural networks train computers to think like humans while digesting "big data" and performing computations quickly, accurately, and without prejudice. Among the many current applications, ML is frequently used to forecast weather, diagnose medical conditions, predict election results, recognize faces, aid the criminal justice system, determine creditworthiness, look at customer traffic, automate traffic signals, target advertising, etc. ML is also used in smart cities.

Over the last 10 years, the main focus of the development of smart cities has been the reduction of traffic congestion. Short communication delays between automobiles and RSUs, efficient traffic flow, and road safety are the key issues with ITSs. Due to the rapid increase in the number of automobiles on the road, traffic congestion and the number of accidents has increased. To solve this issue, vehicle networks (VNs) have developed a number of cutting-edge ideas, including vehicle communications, navigation, and traffic management. ML is a useful technique for discovering hidden insights within ITS without explicit programming since it learns from data. Herein, the research introduced a fusion-based intelligent traffic congestion management system for virtual networks

DOI: 10.1201/9781032648323-1

(FITCCS-VN) that employs ML techniques to collect traffic data and route traffic on available routes in order to lessen traffic congestion in smart cities [13–16]. In order to build "smarter" communities that will enhance the quality of life for people, smart city development requires radical changes in design and the use of cutting-edge technology. The European Commission has launched distinctive smart city projects in four areas: energy, buildings, heating and cooling systems, and transportation. An intelligent transportation system might improve traffic flow in smart cities by monitoring traffic patterns and adjusting the timing of traffic signals. The objective is to promote an ITS that employs real-time data and TMSs to reduce congestion, as well as to identify and encourage ecologically favorable modes of transportation. Applications that utilize less energy, gas, and fuel are considered green. In a smart city, ITS employs cutting-edge and emerging technologies to increase mobility and reduce expenses. The recent advancements in AI, ML, and IST monitor the issues in smart cities like pollution, traffic congestion, and other negative effects. Traffic congestion affects people's quality of life because it intensifies already severe environmental pollution and impairs transportation efficiency. Traffic congestion therefore has an impact on a nation's production, economic growth, and human activities. The best way to handle traffic congestion is the most important issue in urban planning. A lot of research has been done on how to control traffic congestion, and over the last several decades, various initiatives have created a variety of solutions. Intelligent transportation technologies and traffic data collection have grown over time as solutions to these issues [2].

The Internet of Things (IoT), a crucial technology, is particularly beneficial in the development of smart cities. Concerns have been expressed about IoT security and privacy issues, some of which come from a number of sources, including cyberattacks and unsecured networks, data, connections, or communication [3]. The idea is to use the combined knowledge of AI, IoT, and Big Data to address the issues of environmental sustainability and climate change. Smart city technology and solutions are creating a wide range of new possibilities for improved comprehension, communication, and decision-making [4].

Around the world, there have been several initiatives to make cities "smart." Given the broad use of the idea of the smart city in several fields, including urban planning, electrical engineering, and computer sciences, it is challenging to comprehend and define smart cities from various viewpoints [5]. In smart cities, fires may have disastrous effects, resulting in property loss and putting residents' lives at peril. It is difficult to identify fires in real time due to the accuracy and speed constraints of traditional fire detection techniques. The above reference introduced the smart fire detection system (SFDS), an improved fire detection technique for smart cities based on the YOLOv8 algorithm that makes use of the advantages of DL to swiftly identify features specific to fires. The SFDS approach has the potential to improve fire detection accuracy, reduce false alarms, and be more economical than traditional fire detection methods. It might be enhanced to discover more intriguing aspects of smart cities, including gas leaks or floods [6]. Smart surveillance networks help government and law enforcement decision-makers at command centers to be more prepared and effective in handling problems. They consist of a network of nodes and transmitters that carry video and data quickly from one node to another on an ongoing basis. By superimposing real-world and stored data in command center repositories, video, and data analytics systems accelerate the process of analyzing and identifying the resources needed to solve emergencies. As a result, systems for video/data analytics, wireless mesh networks, and fixed and mobile surveillance sensors must all have strong multimodal capabilities [7].

Smart cities refer to the intelligent management of urban areas via the use of numerous electronic sensors and cutting-edge communication techniques. IoT devices and smart city residents are seamlessly connected, and private cloud networks are used to deliver the services. All data gathered from users and devices in the ecosystem is stored in the central cloud database of the smart city infrastructure [8]. Smart people, smart roads, smart parking, smart houses, and smart health are all common components of a smart city. Technology experts, engineers, and developers are currently looking at smart IoT-based solutions to create the ideal structural design for a smart city. The fundamental goals of a smart city are based on three steps: first, gathering data and learning from statistics related to urban environments; second, managing, evaluating, and handling information to carry out independent judgments; and third, service-level implementation of the decided upon course of action [9].

Numerous privacy, social, legal, and ethical concerns arise as a result of LocBigData's increasing popularity and its use in many facets of our everyday lives. What social and ethical difficulties are present in LocBigData research? Although the topic in general has received a lot of attention, these two important questions are still not entirely answered. Also, How can these problems be solved to guarantee the proper use of LocBigData? Both technological and non-technical (regulatory and governance) approaches should be used to answer these questions. There is a pressing need to categorize and model such concerns, as well as to create data analytics methods that are morally conscious and privacy-preserving. Providing "best practices" for resolving such problems and establishing tools to evaluate conformity or discover infractions against correct utilization of LocBigData are also crucial from non-technical viewpoints, such as data governance [10].

Each component involves human resources and funding and is mechanized using various technologies [11]. The key findings from a comprehensive qualitative study and keyword co-occurrence clustering demonstrate that the IoT is used in conjunction with a number of different technologies, including cloud computing, big data analytics, blockchain, AI, and wireless communication networks. Some of the most significant IoT applications for smart cities are smart grids, smart parking, smart buildings, and transportation [12].

1.2 ETHICAL AI

Around 54% of the world's population currently resides in cities, and by 2050, that percentage is projected to rise to 66%, according to recent research. Economic incentives are what propel rapid urbanization, but they also has substantial unintended environmental and social consequences. Therefore, maintaining a balance between the fast increase of urbanization and the resources of the cities depends greatly on environmental, social, and economic sustainability. Modern technology has made it possible to work towards improving urban life's economic, social, and environmental elements while easing the difficulties that come with it. The idea of "smart cities" has recently gained popularity. It promises to employ cutting-edge technology, including a variety of IoT sensors, to gather and analyze data on many aspects of urban life. A collaborative effort between experts from several fields, including engineering, architecture, urban planning, and economics, is required to develop, create, execute, and deploy a smart solution for an underlying job in a smart city application. In order to manage and use resources more effectively, AI approaches have also been shown to be particularly successful in gaining insights from data gathered by various IoT sensors. Healthcare, transportation, education, the environment, agriculture, defense, and public services are some important smart city applications where AI has proven to be highly successful [17].

The utilization of AI-based Internet of Things (IoT) devices for gathering information from individuals underscores the critical importance of addressing data privacy concerns. The collection of personal data raises significant implications for the rights of data subjects to privacy. It imposes limitations on their ability to exercise control over their personal information. As such, it becomes imperative to carefully consider and safeguard data privacy in the deployment and operation of AI-enabled IoT devices. This entails implementing robust privacy policies, incorporating encryption mechanisms, and adhering to regulatory frameworks to ensure the protection of individuals' privacy rights amidst the proliferation of data-driven technologies [21–32]. Once a user has given permission and the personal data has been gathered and processed, the decision regarding privacy preferences and personal information cannot be changed or reversed. The right to privacy does, however, also include the right to change privacy preferences (such as by removing oneself from a model's training data set) or any personal information (such as by changing or updating any individual learned model feature). One should be able to easily and quickly alter their perspective on privacy. Although such assaults may be morally acceptable, they are not harmless and inflict collateral damage to other users. Contrarily, commercially implemented algorithms do not provide means to apply that modification, which is why users have sometimes made a justifiable decision to obfuscate using an adversarial system [19].

1.3 APPLICATIONS OF ETHICAL AI IN SMART CITIES

A smart city is expected to be composed of safe, secure, and trustworthy connected sensors, actuators, and relays in order to gather, analyze, and send data to give reliable and efficient digital services. The vast majority of the data is generated by cloud-based IoT devices, which are essential to many applications used in smart cities. Researchers have studies the importance of the preservation of data security and privacy, the defense of networks against future cyberattacks, the encouragement of an adult and responsible data-sharing culture, and the employment of AI and ML techniques in real-world

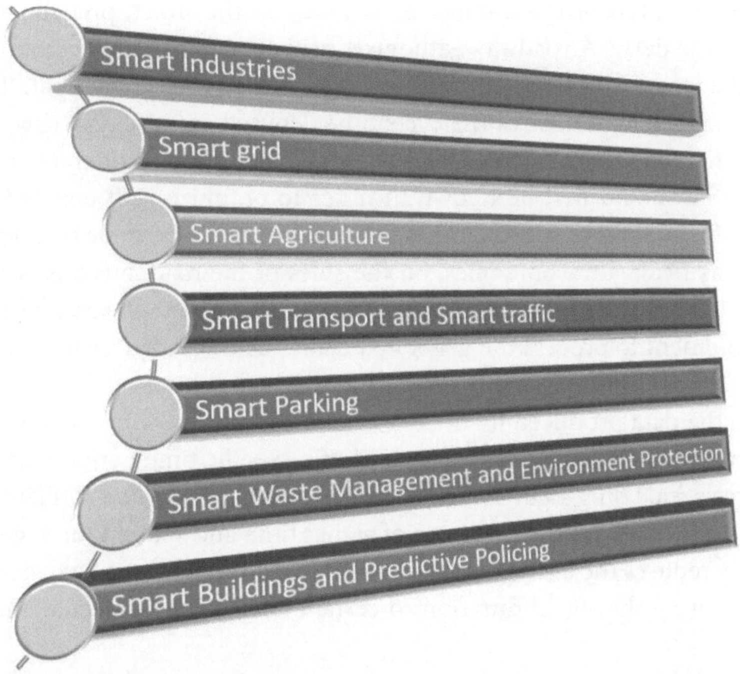

FIGURE 1.1 ML Applications in Smart Cities.

applications such as smart cities [20] (Figure 1.1). In the following sections we discuss different aspects of smart cities.

1.4 SMART TRANSPORT

A technologically sophisticated metropolitan area known as a "smart city" has a number of cutting-edge features built on information and communication technologies (ICT) and other electronic methods. One of a smart city's distinguishing characteristics is its smart transportation system, which encourages effective vehicle mobility and provides people with convenient, secure transportation. The two primary causes of collisions and incidents on city streets are smog and severe air pollution. Smog, a mixture of smoke and fog, is created by the ozone (O3) and particulate matter, which includes pollen, dust, sulfur oxides, and other pollutants. If the air contaminants and their quantity are known, we may thus rapidly determine if smog is present in the air. Different regression models, such as the Polynomial Regression Model, Decision Tree Regression Model, Random Forest Regression Model, and Support Vector Regression Model are used to predict the Air Quality Index (AQI), from which the pollutant concentration can be determined, which will help to detect the smog [33].

1.4.1 Smart Traffic

Current traffic systems employ fixed signal timings at traffic signal intersections, which is the primary cause of traffic congestion. Because the same amount of time is allotted for each lane regardless of the number of vehicles on the street, predefined set signal times increase the delay. A roadway with fewer vehicles can pass in a certain amount of time, while the lane with more vehicles must wait for another green light. Therefore, it is necessary to make better use of the cycle time. Unused excess signal time is not used properly. The first and most important phase, traffic detection, is often overlooked. The amount of traffic should first be known in order to optimize it. It entails figuring out how much traffic there is at specific times or intervals. It can be done by putting several sensors, such as IR or ultrasonic ones, on the sides of the road, but this is not a smart idea because it necessitates a lot of administration, uses more power, and needs more advanced equipment to process the generated data. The usage of a security camera is an additional method. Using image processing and methods, object recognition algorithms may examine the data produced by the camera installed at the intersection. The goal of traffic prediction is to forecast traffic in a lane at a specific time period using historical data for a given area. Using algorithms, this traffic estimate is created. Utilizing resources effectively includes making the most use of signal time and space. Our suggested solution effectively reduces the average traveler's wait at a given junction by dynamically and efficiently allocating the signal duration to certain lanes using projected and historical data [34].

A key issue in the transport system is the enormous rise in traffic load and congestion brought on by the large number of private and public cars during the past two decades. An intelligent traffic management system becomes an essential part of the transportation system in order to efficiently manage the traffic in smart cities. The authors of Ref. [18] suggested an intelligent traffic control system in this research to create a smart platform in Mandalay, Myanmar with the goals to cut down on traffic jams, car accidents, fuel use, and travel time. This study analyses the volume of traffic data using big data technologies and, and forecasts the ideal road traffic used to deliver safe, comfortable, and less stressful travel. A collection of tools and services that are connected to different modes of transportation and traffic control software make up an ITS. The main purposes of these systems are to provide people with thorough transportation information, shorten commutes (travel times), and ease traffic congestion. As a result, the accident rate will decrease as well. Intelligent traffic management systems may be utilized to reduce crashes and traffic congestion while also achieving specified targets for commuter comfort.

1.4.2 Smart Parking

Finding available parking spots in large parking lots can be a significant source of time wastage. Addressing this issue, this research introduces a novel approach aimed at minimizing the time lost in the search for parking spaces. The proposed solution is dubbed 'Smart Parking-Lots Based on AI Techniques' (SPLAI). SPLAI serves as an intelligent parking system that leverages advanced AI methods to make decisions about granting access to both users and vehicles entering the parking facility.

In operation, drivers bring their vehicles to a designated platform within the parking area. At this point, the system takes over, handling the process of parking. Notably, the study shows that parked vehicles are inclined at a gentle angle, with a maximum of 30 degrees. This inclination serves the purpose of preventing any potential contact between the vehicle's undercarriage and the ground. In essence, SPLAI's core objective is to transform the conventional parking experience by using AI-driven strategies to enhance parking space availability. By seamlessly guiding vehicles to open parking spots and employing a systematic approach to admission, the system seeks to minimize the considerable time often spent on searching for parking spaces. This approach also permits the accommodation of more cars within the same space, leading to a remarkable improvement in parking lot efficiency. This optimization can lead to an astounding 40% reduction in parking lot space requirements.

1.4.3 Smart Waste Management

A significant question that merits in-depth academic examination is how smart cities could be able to handle more general environmental challenges, particularly trash management. This research explores the potential for smart cities and connected communities to support more successful waste management efforts by drawing on an integrated evaluation of the available literature. The shortcomings of the existing waste management methods are highlighted, and a conceptual framework for a centralized waste management system is proposed. This framework is made up of three related parts: 1) Infrastructure for Holistic Product Lifespan Data Collection: The goal of this component is to provide a solid infrastructure that will allow for the collection of thorough data throughout the course of a product's full lifespan. The optimization of waste management operations depends on the visibility of the data. 2) unique Business Models Based on Product lifetime Information: A variety of unique business models are investigated, all of which are based on product lifetime information. This strategy seeks to reduce trash production, encouraging sustainable activities. 3) Intelligent Sensor-Based Infrastructure for Upstream Waste Separation and Collection: In this component, sensors are integrated into an intelligent infrastructure to enable effective waste separation at the source and prompt collection. The suggested paradigm emphasizes how crucial product lifecycle data is to reducing waste production and fostering trash recovery. Additionally, it emphasizes how important it is to match waste management plans with a product's whole lifespan. The story also explores how monitoring and data-sharing technologies are used in practice to look into waste management issues. At the conclusion of the discussion, the chapter zeroes in on the key components vital for effectively executing the proposed framework within the context of smart cities. It also explores promising directions for ongoing research endeavors in this dynamic field [35–40]. Emphasizing clarity and coherence, the chapter underscores the necessity of integrating diverse technological and socio-economic factors to address the unique challenges and opportunities of urban environments. By fostering interdisciplinary collaborations and stakeholder engagement, it aims to drive sustainable and inclusive urban development [41–44]. Through this approach, the discussion endeavors to catalyze positive transformations in urban living, leveraging the potential of smart city initiatives to enhance quality of life and address pressing societal needs. [40].

1.4.4 Smart Buildings

In today's rapidly changing landscape, adaptability, resilience, and sustainability have taken center stage as the key concepts driving innovation. We are witnessing the creation of structures that not only cater to the needs of people using them but also establish seamless channels for gathering real-time data, yielding valuable insights. A recent study has turned the spotlight on an exciting possibility: by harnessing the power of intelligent technologies, we could potentially slash urban carbon emissions by a staggering 90%. This reduction would encompass cities, their architectural marvels, and the underlying infrastructure. For buildings specifically, the focus lies in devising solutions that enhance the efficiency of renewable energy systems, optimize water resource management, improve air quality, and encourage the widespread adoption of environmentally conscious practices

Leading experts, like Gartner, paint a vivid picture of the future, where commercial smart buildings will house a staggering four billion IoT devices by the year 2028. These devices will harness the prowess of telecommunications infrastructures, taking advantage of the blazing speeds of 5G and the hyper-efficient connectivity of 6 or 6E Wi-Fi. Moreover, they'll be equipped with smart utilities meticulously designed to meet operational demands.

1.5 CONCLUSION

As people rely more and more on these so-called "smart machines" for anything from easy activities to crucial life decisions, ML the core component of AI, is quickly transforming how people live their lives. The elimination of traffic congestion has been the primary goal of the creation of smart cities for the last ten years. The smooth flow of traffic, road safety, and brief communication delays between vehicles and RSUs are the key issues with ITSs. Traffic congestion and the number of accidents have grown as a result of the rapid growth in the number of vehicles on the road. Smart cities have been built utilizing to lessen these problems. We have addressed Ethical ML, its framework, and applications of smart cities using ethical ML in this article, which is the investigation and analysis of smart cities using ML from many research publications. Ethical gives a brief idea about it and the framework of ML is explained in detail. Additionally, we have explored the many ML-based applications for smart cities, including smart waste management, smart buildings, smart parking, smart agriculture, smart traffic, and smart transportation. Here we can conclude that techniques of IoT and blockchain play a vital role along with ML techniques.

BIBLIOGRAPHY

[1] C. Malhotra, "Ethical Framework for Charru Malhotra," *Mach. Learn. a 5G Futur.*, pp. 26–28, 2018.

[2] M. Saleem, S. Abbas, T. M. Ghazal, M. Adnan Khan, N. Sahawneh, and M. Ahmad, "Smart cities: Fusion-based intelligent traffic congestion control system for vehicular networks

using techniques," *Egypt. Informatics J.*, vol. 23, no. 3, pp. 417–426, 2022, doi: 10.1016/j.eij.2022.03.003.

[3] E. M. Onyema, S. Dalal, C. A. T. Romero, B. Seth, P. Young, and M. A. Wajid, "Design of intrusion detection system based on cyborg intelligence for security of cloud network traffic of smart cities," *J. Cloud Comput.*, vol. 11, no. 1, 2022, doi: 10.1186/s13677-022-00305-6.

[4] S. E. Bibri, A. Alexandre, A. Sharifi, and J. Krogstie, "Environmentally sustainable smart cities and their converging AI, IoT, and big data technologies and solutions: an integrated approach to an extensive literature review," *Energy Informatics*, vol. 6, no. 1, 2023, doi: 10.1186/s42162-023-00259-2.

[5] C. Lim, G. H. Cho, and J. Kim, "Understanding the linkages of smart-city technologies and applications: Key lessons from a text mining approach and a call for future research," *Technol. Forecast. Soc. Change*, vol. 170, no. September 2020, p. 120893, 2021, doi: 10.1016/j.techfore.2021.120893.

[6] F. M. Talaat and H. ZainEldin, "An improved fire detection approach based on YOLO-v8 for smart cities," *Neural Comput. Appl.*, vol. 1, no. 35, 2023, doi: 10.1007/s00521-023-08809-1.

[7] M. Kashef, A. Visvizi, and O. Troisi, "Smart city as a smart service system: Human-computer interaction and smart city surveillance systems," *Comput. Human Behav.*, vol. 124, no. May, p. 106923, 2021, doi: 10.1016/j.chb.2021.106923.

[8] N. A. Megahed and R. F. Abdel-Kader, "Smart cities after COVID-19: Building a conceptual framework through a multidisciplinary perspective," *Sci. African*, vol. 17, no. 1, p. e01374, 2022, doi 10.1016/j.sciaf.2022.e01374.

[9] A. Ullah *et al.*, "Smart Cities: The Role of Internet of Things and in Realizing a Data-Centric Smart Environment," *Complex Intell. Syst.*, pp. 1–31, 2023, doi: 10.1007/s40747-023-01175-4.

[10] H. Huang, X. A. Yao, J. M. Krisp, and B. Jiang, "Analytics of location-based big data for smart cities: Opportunities, challenges, and future directions," *Comput. Environ. Urban Syst.*, vol. 90, no. September, p. 101712, 2021, doi 10.1016/j.compenvurbsys.2021.101712.

[11] R. Al-Sharif and S. Pokharel, "Smart City Dimensions and Associated Risks: Review of literature," *Sustain. Cities Soc.*, vol. 77, no. 1, p. 103542, 2022, doi: 10.1016/j.scs.2021.103542.

[12] A. Rejeb, K. Rejeb, S. Simske, H. Treiblmaier, and S. Zailani, "The big picture on the Internet of things and the smart city: A review of what we know and what we need to know," *Internet of Things (Netherlands)*, vol. 19, no. July, p. 100565, 2022, doi: 10.1016/j.iot.2022.100565.

[13] S. Lo Piano, "Ethical principles in and artificial intelligence: Cases from the field and possible ways forward," *Humanit. Soc. Sci. Commun.*, vol. 7, no. 1, pp. 1–7, 2020, doi: 10.1057/s41599-020-0501-9.

[14] A. Kumar *et al.*, "Development of a cloud-assisted classification technique for the preservation of secure data storage in smart cities," *J. Cloud Comput.*, vol. 12, no. 1, 2023, doi: 10.1186/s13677-023-00469-9.

[15] Y. Jiang, T. H. Tran, and L. Williams, "And mixed reality for smart aviation: Applications and challenges," *J. Air Transp. Manag.*, vol. 111, no. May, p. 102437, 2023, doi: 10.1016/j.jairtraman.2023.102437.

[16] G. Mageswaran, S. D. Nagappan, N. Hamzah, and S. N. Brohi, ": An Ethical, Social & Political Perspective," *Proc. - 2018 4th Int. Conf. Adv. Comput. Commun. Autom. ICACCA 2018*, pp. 1–6, 2018, doi: 10.1109/ICACCAF.2018.8776702.

[17] K. Ahmad, M. Maabreh, M. Ghaly, K. Khan, J. Qadir, and A. Al-Fuqaha, "Developing Future Human-Centered Smart Cities: Critical Analysis of Smart City Security, Interpretability, and Ethical Challenges," 2020, [Online]. Available: http://arxiv.org/abs/2012.09110

[18] S. S. Hlaing, M. M. Tin, M. M. Khin, P. P. Wai, and G. R. Sinha, "Big traffic data analytics for smart urban intelligent traffic system using techniques," *2020 IEEE 9th Glob. Conf. Consum. Electron. GCCE 2020*, pp. 299–300, 2020, doi: 10.1109/GCCE50665.2020.9291790.

[19] L. Adomaitis and R. Oak, "Ethics of adversarial and data poisoning," *Digit. Soc.*, vol. 2, no. 1, pp. 1–13, 2023, doi: 10.1007/s44206-023-00039-1.

[20] Z. Ullah, F. Al-Turjman, L. Mostarda, and R. Gagliardi, "Applications of artificial intelligence and in smart cities," *Comput. Commun.*, vol. 154, no. February, pp. 313–323, 2020, doi: 10.1016/j.comcom.2020.02.069.

[21] S. Pandya *et al.*, "Federated learning for smart cities: A comprehensive survey," *Sustain. Energy Technol. Assessments*, vol. 55, no. December 2022, p. 102987, 2023, doi 10.1016/j.seta.2022.102987.

[22] Z. Wang, W. Cheng, and C. Li, "DoS Attack Detection Model of Smart Grid based on Method," *Proc. 2020 IEEE Int. Conf. Power, Intell. Comput. Syst. ICPICS 2020*, pp. 735–738, 2020, doi 10.1109/ICPICS50287.2020.9202401.

[23] J. Hao *et al.*, "An adaptive markov strategy for defending smart grid false data injection from malicious attackers," *IEEE Trans. Smart Grid*, vol. 9, no. 4, pp. 2398–2408, 2018, doi: 10.1109/TSG.2016.2610582.

[24] R. Rituraj, D. Ecker, and V. K. Annamaria, "Data-Driven Methods of in modeling the Smart Grids," *SISY 2022 - IEEE 20th Jubil. Int. Symp. Intell. Syst. Informatics, Proc.*, pp. 89–96, 2022, doi: 10.1109/SISY56759.2022.10036268.

[25] D. Oreski, I. Pihir, and K. Cajzek, "Smart Agriculture and Digital Transformation on Case of Intelligent System for Wine Quality Prediction," *2021 44th Int. Conv. Information, Commun. Electron. Technol. MIPRO 2021 - Proc.*, pp. 1370–1375, 2021, doi: 10.23919/MIPRO52101.2021.9596979.

[26] J. Treboux, R. Ingold, and D. Genoud, "Towards Retraining of Algorithms: An Efficiency Analysis Applied to Smart Agriculture," *GIoTS 2020 - Glob. Internet Things Summit, Proc.*, pp. 2–7, 2020, doi: 10.1109/GIOTS49054.2020.9119601.

[27] S. Prince Samuel, K. Malarvizhi, S. Karthik, and S. G. Mangala Gowri, "Internet of Things based Smart Agriculture," *2020 6th Int. Conf. Adv. Comput. Commun. Syst. ICACCS 2020*, pp. 1101–1106, 2020, doi 10.1109/ICACCS48705.2020.9074472.

[28] F. Raimundo, A. Gloria, and P. Sebastiao, "Prediction of Weather Forecast for Smart Agriculture supported by," *2021 IEEE World AI IoT Congr. AIIoT 2021*, pp. 160–164, 2021, doi: 10.1109/AIIoT52608.2021.9454184.

[29] G. Abraham, R. Raksha, and M. Nithya, "Smart Agriculture Based on IoT and," *Proc. - 5th Int. Conf. Comput. Methodol. Commun. ICCMC 2021*, no. Iccmc, pp. 414–419, 2021, doi 10.1109/ICCMC51019.2021.9418392.

[30] G. Venkatakotireddy, C. Lakshminatha Reddy, R. Ramyadevi, J. Prabhakaran, T. Nivethitha, and R. Mahaveerakannan, "Development of High-Quality Crops using Optimized in Smart Agriculture Environment," *Proc. 3rd Int. Conf. Artif. Intell. Smart Energy, ICAIS 2023*, no. Icais, pp. 532–536, 2023, doi: 10.1109/ICAIS56108.2023.10073727.

[31] H. M. K. K. M. B. Herath and M. Mittal, "Adoption of artificial intelligence in smart cities: A comprehensive review," *Int. J. Inf. Manag. Data Insights*, vol. 2, no. 1, p. 100076, 2022, doi: 10.1016/j.jjimei.2022.100076.

[32] U. Ahmad, H. Song, A. Bilal, M. Alazab, and A. Jolfaei, "Securing smart vehicles from relay attacks using," *J. Supercomput.*, vol. 76, no. 4, pp. 2665–2682, 2020, doi: 10.1007/s11227-019-03049-4.

[33] A. Mishra, Z. M. Jalaluddin, and C. V. Mahamuni, "Air Quality Analysis and Smog Detection in Smart Cities for Safer Transport using ML Regression Models," *Proc. - 2022 IEEE 11th Int. Conf. Commun. Syst. Netw. Technol. CSNT 2022*, no. Ml, pp. 200–206, 2022, doi: 10.1109/CSNT54456.2022.9787618.

[34] S. C. Kanigolla, "Smart Traffic Detection and Control using Techniques," *2023 5th Int. Conf. Energy, Power Environ. Towar. Flex. Green Energy Technol.*, pp. 1–4, 2023, doi: 10.1109/ICEPE57949.2023.10201549.

[35] F. H. Alshamrani, H. F. Syed, and M. A. Elhussein, "Machine learning based model for traffic prediction in smart cities," *IET Conf. Publ.*, vol. 2019, no. CP758, pp. 4–9, 2019, doi: 10.1049/cp.2019.0195.

[36] S. Chuanxia, Z. Han, and Y. Peixuan, "Machine learning and and IoTs for forecasting prediction of smart road traffic flow," *Soft Comput.*, vol. 27, no. 1, pp. 323–335, 2023, doi: 10.1007/s00500-022-07618-3.

[37] W. H. F. Aly, Y. El Masry, and O. Al-Khatlan, "Smart Parking-Lots Based on AI Techniques SPLAI," *BioSMART 2023 - Proc. 5th Int. Conf. Bio-Engineering Smart Technol.*, pp. 1–5, 2023, doi: 10.1109/BioSMART58455.2023.10162123.

[38] M. Jemmali, L. K. B. Melhim, M. T. Alharbi, A. Bajahzar, and M. N. Omri, "Smart-parking management algorithms in smart city," *Sci. Rep.*, vol. 12, no. 1, pp. 1–15, 2022, doi: 10.1038/s41598-022-10076-4.

[39] H. Umer, A. Ahmed, F. Ali, S. Sarmad Ali, and M. Ali Khan, "A Systematic Literature Review on Smart Waste Management Using," *Proc. 2022 Mohammad Ali Jinnah Univ. Int. Conf. Comput. MAJICC 2022*, pp. 1–9, 2022, doi: 10.1109/MAJICC56935.2022.9994104.

[40] B. Esmaeilian, B. Wang, K. Lewis, F. Duarte, C. Ratti, and S. Behdad, "The future of waste management in smart and sustainable cities: A review and concept paper," *Waste Manag.*, vol. 81, pp. 177–195, 2018, doi: 10.1016/j.wasman.2018.09.047.

[41] L. chu Su, X. Wu, X. Zhang, and X. Huang, "Smart performance-based design for building fire safety: Prediction of smoke motion via AI," *J. Build. Eng.*, vol. 43, no. April, p. 102529, 2021, doi: 10.1016/j.jobe.2021.102529.

[42] V. Bijlani, "Smart Buildings for Sustainable Smart Cities," *1st Int. Conf. Adv. Innov. Smart City, ICAISC 2023 - Proc.*, pp. 1–6, 2023, doi: 10.1109/ICAISC56366.2023.10085629.

[43] D. Correia, L. Teixeira, and J. L. Marques, "Study and analysis of the relationship between smart cities and Industry 4.0: A systematic literature review," *Int. J. Technol. Manag. Sustain. Dev.*, vol. 21, no. 1, pp. 37–66, 2022, doi: 10.1386/tmsd_00051_1.

[44] L. U. Khan, M. Alsenwi, I. Yaqoob, M. Imran, Z. Han, and C. S. Hong, "Resource optimized federated learning-enabled cognitive internet of things for smart industries," *IEEE Access*, vol. 8, no. 1, pp. 168854–168864, 2020, doi: 10.1109/ACCESS.2020.3023940.

CHAPTER 2

Ethical Machine Learning for Internet of Things Network

Maysa Khalil and Qasem Abu Al-Haija

2.1 INTRODUCTION

Connecting normal devices to the Internet, known as the Internet of Things (IoT), has unexpectedly grown and is expected to continue increasing. By 2025, there will be 75 billion connected devices, as noted by Jagadeesan et al. [1]. This increase in IoT devices has led to concerns regarding privacy and protection and other ethical issues. IoT devices have the potential to gather sensitive data such as location data, which makes them vulnerable to cyberattacks and hacking. Therefore, it is essential to protect both the security and privacy of IoT devices as well as the privacy of their data.

An IoT system's connected devices can be monitored, managed, and used with artificial intelligence (AI) and machine learning (ML). AI plays a pivotal role in integrating numerous IoT devices onto a unified platform, enabling centralized control with a single touch. This consolidation facilitates seamless management and coordination of diverse devices. Additionally, AI empowers the exploration of edge and cloud computing capabilities within this platform, unlocking potential for enhanced efficiency and scalability in IoT operations. Through AI-driven integration and computing advancements, the possibilities for optimizing IoT functionality and connectivity continue to expand, promising greater convenience and flexibility in device management. Figure 2.1 shows the IoT, big data, AI, and ML/DL cycle. The billions of internet-connected devices spread across the globe create data, which is saved online and acts as a collective memory for training ML algorithms.

Gorodishcheva et al. [2] discuss ethical concerns, the importance of responsible AI practices, and the need for a framework for integrating ethical considerations into the

DOI: 10.1201/9781032648323-2

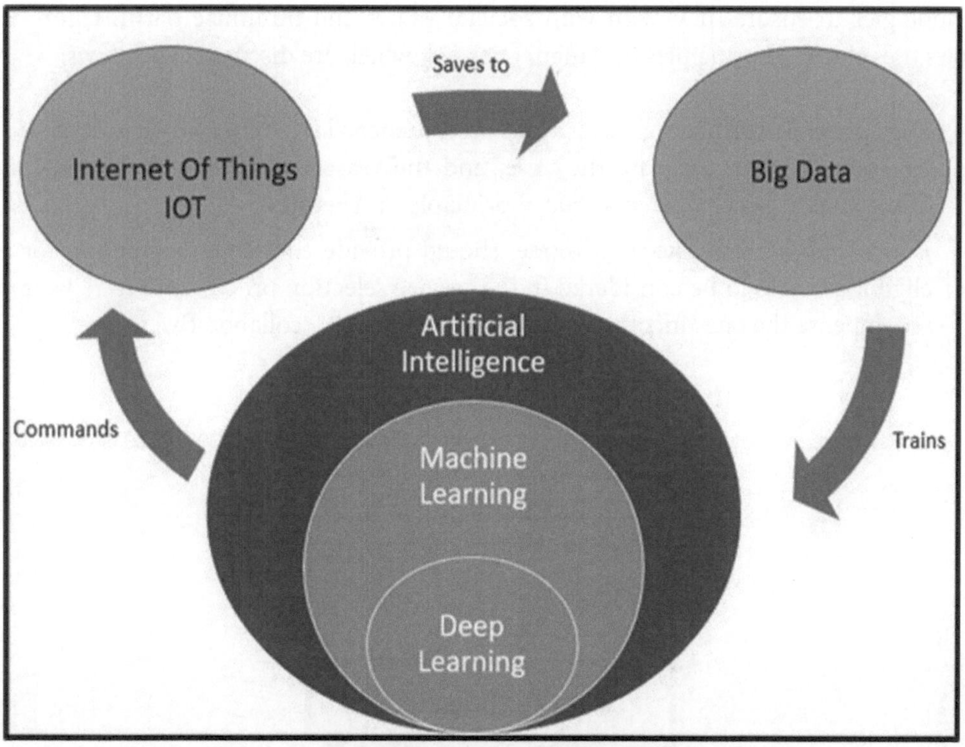

FIGURE 2.1 Relationship between IoT, Big Data, AI, Machine and Deep Learning.

development and deployment of ML models for IoT networks. They emphasize the need for robust data governance, privacy-preserving techniques, and algorithmic fairness to ensure that IoT systems not only deliver innovative solutions but also uphold the rights and dignity of individuals using the devices.

ML techniques, particularly deep learning algorithms, are developed incrementally to enhance performance on a specific task. ML algorithms are commonly assessed based on their accuracy when applied to a particular dataset or task, such as segmentation or classification. Task performance comparisons are now widely accepted within the ML community. In the case of classification tasks, commonly used metrics for evaluation include accuracy, precision, recall, receiver operating characteristics (ROC), and the area under the curve (AUC), but ethical concerns are not used to evaluate the ML performance.

The role of regulatory bodies, industry standards, and ethical guidelines in shaping the landscape of ethical ML within IoT networks should be considered. With the help of the abovementioned technologies ethical use of technologies like IoT, AI. and ML is made possible.

2.2 AI ETHICAL PRINCIPLES

AI ethical principles provide a framework for the responsible development and use of AI systems. These principles help guide the design, deployment, and governance of AI technologies to ensure they align with societal values and minimize harm. Figure 2.2 shows the ethical AI principles in Bangui et al. [3], which are discussed as follows:

- *Transparency*: Providing clear and comprehensible information about how AI systems work, the data they use, and the reasoning behind their decisions. Transparency is considered a crucial principle in AI ethics.
- *Justice and fairness*: The trust model should provide equitable possibilities for all eligible trustees to be considered in the service selection process and to redeem or recompense the ones incorrectly assessed in the social, collaborative setting.

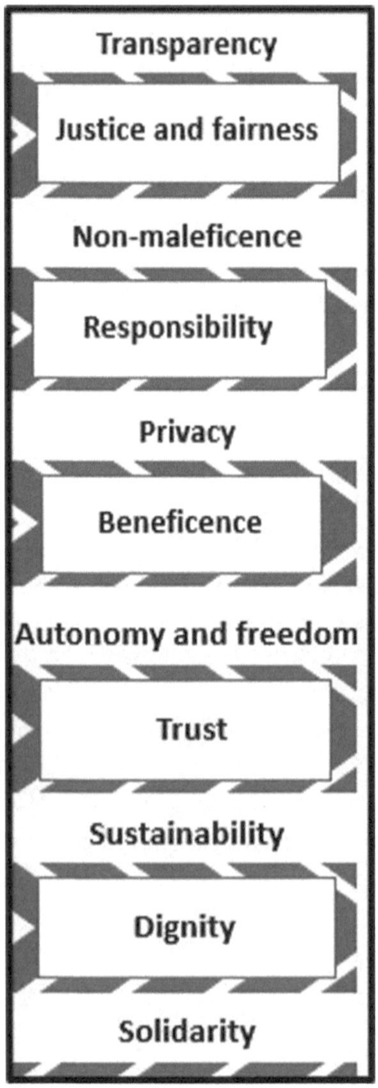

FIGURE 2.2 AI Ethical Principles.

- *Non-maleficence*: The harmful items that may disturb the reputation of the participants in the social, collaborative environment or harm them should be actively eliminated by the trust model.
- *Responsibility*: The trust model should encourage the gathering of data that prevents harmful participants from evading responsibility for their misdeeds in a socially collaborative setting by challenging their accountability.
- *Privacy:* The social collaborative environment's members' privacy should be protected, and the trust model should be able to tell the difference between public and private information.
- *Beneficence:* As a sign of well-being, the trust model should guarantee peaceful social collaboration and co-creation to enhance human life.
- *Autonomy and freedom:* Even though it would enable the social, collaborative environment to optimize a certain global goal such as addressing water availability issues, the trust model should respect people's autonomy and freedom and recognize that there is a line that should not be passed.
- *Trust*: The trust model should create and maintain reliable connections between intelligent things.
- *Sustainability:* To support the continuous participation of all members in productive and mutually beneficial collaboration, the trust model should strike a balance between the common and individual members' special interests within the social, collaborative environment.
- *Dignity:* To enable socialization in an IoT society, the trust model should guarantee that human dignity will be upheld and protected.
- *Solidarity:* The trust model should enable mechanisms (such as incentive programs) that encourage robust and healthy ecosystem members to help weaker ones without endangering social cohesiveness.

Some ML algorithms consider these AI ethical principles, but not all. Some applications also satisfy some AI ethical principles, but not all. For example, most ML algorithms satisfy Non-maleficence and Beneficence, whereas none satisfy Transparency in some applications [5–9]. However, algorithms like decision trees, linear models, and rule-based systems are often more interpretable than complex deep learning models like neural networks.

Transparency has emerged as a significant ethical concern across the various ML algorithms. hat statement was intended to be more general and did not refer to a specific study or application. It aimed to highlight a common challenge in trust models that utilize AI algorithms: the oversight of potential biases. In many cases, while these models can accurately assess trustworthiness, they may not adequately address biases inherent in the data or algorithms used. This oversight can undermine the fairness and reliability of the trust assessment process [36]. If transparency is lacking, acknowledging the inferences and judgments made by trust models employing AI algorithms is insufficient in guaranteeing absolute fairness, consistency, and trustworthiness. Moreover, concepts such as sustainability, dignity, and solidarity are intricately connected, whether explicitly or implicitly, to the transparency of trust models, as they are influenced by the inherent logic governing these models.

2.3 RELEVANCE BETWEEN INDUSTRIES AND AI/ML

Regulatory and ethical considerations can impact the adoption and deployment of AI/ML in different sectors. The degree of relevance of AI/ML in each industry or sector can vary widely based on several factors, including the industry's specific needs, technological advancements, and regulatory considerations. Here's a general overview of the degree of relevance AI/ML has in different industries, as shown in Figure 2.3.

1. Healthcare: High relevance - AI/ML is used for disease diagnosis, drug discovery, patient monitoring, and personalized treatment plans [10].
2. Finance: High relevance - AI/ML is applied in fraud detection, risk assessment, algorithmic trading, and customer service [11].
3. Retail: High relevance - AI/ML is used for demand forecasting, inventory management, personalized marketing, and chatbots for customer support [12].
4. Manufacturing: High relevance - AI/ML is used for process optimization, predictive maintenance, quality control, and supply chain management [13]
5. Transportation: High relevance - AI/ML is employed in autonomous vehicles, route optimization, traffic management, and predictive maintenance for fleets [14].
6. Agriculture: High relevance - AI/ML is used for precision farming, crop monitoring, pest control, and yield prediction [15].
7. Energy: High relevance - AI/ML is applied in energy grid management, predictive maintenance for power plants, and renewable energy optimization [16][17].
8. E-commerce: High relevance - AI/ML powers product recommendations, fraud detection, and chatbots in online shopping [18].

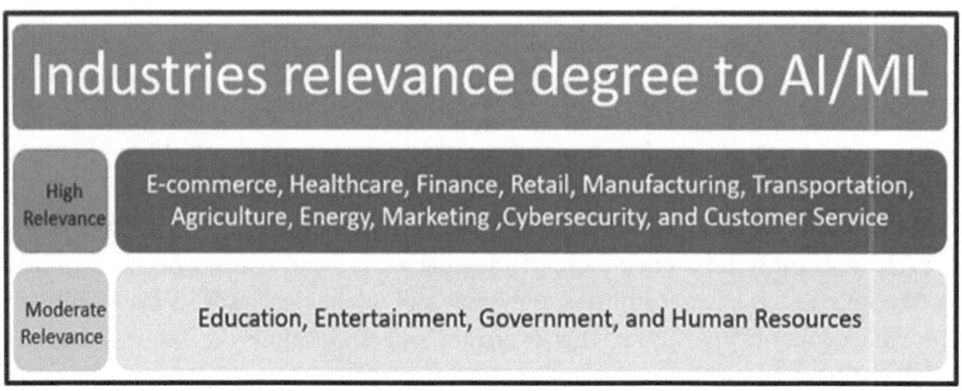

FIGURE 2.3 Industries Relevance Degree to AI/ML.

9. Marketing: High relevance – AI/ML is used for customer segmentation, A/B testing, ad targeting, and content optimization [19].
10. Cybersecurity: High relevance – AI/ML is employed for threat detection, anomaly detection, and identifying security vulnerabilities [20].
11. Customer Service: High relevance – AI/ML-driven chatbots and virtual assistants handle customer inquiries [21].
12. Education: Moderate relevance – AI/ML is used for personalized learning, adaptive assessments, and administrative tasks, but its adoption is still growing.
13. Entertainment: Moderate relevance – AI/ML is used for content recommendation, videogame design, and content creation tools.
14. Government: Moderate relevance – AI/ML is used in public safety, data analysis, and administrative processes, but its adoption varies by region.
15. Human Resources: Moderate relevance – AI/ML is used for talent acquisition, employee engagement, and HR analytics.

The degree of relevance can change over time as AI/ML technologies advance and industries adapt to new possibilities. Some industries may also have subsectors with varying AI/ML adoption degrees.

2.4 IMPORTANCE OF APPLYING AI ETHICAL PRINCIPLES

Applying all AI ethical principles is critically important to open the opportunities to use AI/ML in all fields for the following reasons:

- The AI/ML will be used comfortably because these ethical principles help ensure that AI systems do not cause harm to individuals, communities, or society as a whole. This is particularly vital in sectors like healthcare and autonomous vehicles, where AI decisions can directly impact human lives.
- Applying AI ethical principles across industries is essential for promoting fairness, transparency, and trust building.
- Financially, it has been estimated that AI might boost the world economy by over 16 trillion USD by 2030 [22].
- Creating new systems to serve different and new fields (e.g., Design and Application Approaches in Women-Safety Systems [23].
- Ethical behavior can enhance an organization's reputation and brand value. Companies that demonstrate ethical AI practices are likelier to attract customers and investors who prioritize responsible technology.
- Adhering to ethical AI principles can help organizations avoid legal and regulatory issues. Non-compliance with ethical guidelines may lead to legal consequences, fines, or damage to an organization's reputation.

- Ethical AI practices contribute to the long-term viability of AI solutions. Unethical behavior can lead to public backlash, regulatory action, and economic losses, undermining the sustainability of AI initiatives.
- Ethical AI principles encourage responsible innovation and collaboration among industry players. When industries collectively adhere to ethical guidelines, they can work together to address common challenges and advance technology responsibly.

Therefore, it is imperative to integrate explainable AI ethical principles into the various components of ML models to ensure transparency and minimize the risk of biased decision-making, ultimately paving the way for a standardized AI-based trust model that upholds all 11 ethical dimensions.

2.5 CONCLUSIONS

The choice of specific ML algorithms should be guided by the ethical considerations relevant to the required application. However, ethical AI is not just about the algorithms but also the entire AI development pipeline, data, and decision-making processes. Adopting a holistic approach that considers ethics at every AI development and deployment stage is essential. Additionally, ethical guidelines and regulations may vary by industry and jurisdiction, so staying informed about the specific requirements in specific domains is crucial.

REFERENCES

[1] D. Jagadeesan, R. Sundar, S. Sreekanth, S. Theetchenya, and B. Anandraj, "Real World Autonomous IoT Based Data Privacy Protection Using Machine Learning," 2023 Eighth International Conference on Science Technology Engineering and Mathematics (ICONSTEM), Chennai, India, 2023, pp. 1–8, doi: 10.1109/ICONSTEM56934.2023.10142934.

[2] A. N. Gorodishcheva, Y. V. Paskhalskaya, A. V. Gorodishchev, A. I. Vinogradova and G. P. Kovalev, "IoT Ethic In Scientific Communications," *2021 Communication Strategies in Digital Society Seminar (ComSDS)*, St. Petersburg, Russia, 2021, pp. 136–140, doi: 10.1109/ComSDS52473.2021.9422860.

[3] H. Bangui, B. Buhnova, and M. Ge, "Social Internet of Things: Ethical AI Principles in Trust Management," *Procedia Comput. Sci.*, vol. 220, C 2023, pp. 553–560, https://doi.org/10.1016/j.procs.2023.03.070

[4] C. Zhang, W. Li, Y. Luo, and Y. Hu, "AIT: An AI-Enabled Trust Management System for Vehicular Networks Using Blockchain Technology," in *IEEE Internet of Things Journal*, vol. 8, no. 5, 1 March, 2021, pp. 3157–3169, doi: 10.1109/JIOT.2020.3044296.

[5] P. Hankare, S. Babar, P. i Mahalle, Trust Management Approach for Detection of Malicious Devices in SIoT. *Tehnički Glasnik*, vol. 15, no. 1, 2021, pp. 43–50, https://doi.org/10.31803/tg-20210204180217

[6] R. Magdich, H. Jemal, and M. Ben Ayed, "Context-Awareness Trust Management Model for Trustworthy Communications in the Social Internet Of Things." *Neural Comput. Appl. 34*, 24, Dec 2022, pp. 21961–21986, https://doi.org/10.1007/s00521-022-07656-w

[7] S. Sagar, M. Kai Wang, Q. Z. Sheng, and W. E. Zhang. "Trust-SIoT: Towards Trustworthy Object Classification in the Social Internet of Things." *ArXiv*, 2022. Accessed September 23, 2023. /abs/2205.03226.

[8] A. Meena Kowshalya, and M. L. Valarmathi, "Trust Management for Reliable Decision Making among Social Objects in the Social Internet of Things,". *IET Netw.*, vol. 6, 2017, pp. 75–80, https://doi.org/10.1049/iet-net.2017.0021

[9] K. Albulayhi, Q. A. Al-Haija, Q. A.; Sheldon, F. T., 'Effective learning-based attack detection methods for the Internet of Things', 'AIoT Technologies and Applications for Smart Environments,' Chap. 15, Computing, 2022, pp. 275–293, DOI: 10.1049/PBPC057E_ch15, IET Digital Library.

[10] S. Nandi, M. Mishra, S. Majumder, "Usage of AI and Wearable IoT Devices for Healthcare Data: A Study," in *Machine Learning Algorithms for Signal and Image Processing*, IEEE, 2023, pp.315–337, doi: 10.1002/9781119861850.ch18.

[11] S. Gupta, "Role of Internet of Things (IoT) in Smart Finance and Banking," *2023 International Conference on Computational Intelligence, Communication Technology and Networking (CICTN)*, Ghaziabad, India, 2023, pp. 467–470, doi: 10.1109/CICTN57981.2023.10140915.

[12] O. Gupta and P. Joshi, "Applications of IoT in Retail: A Systematic Review," *2023 International Conference on Device Intelligence, Computing and Communication Technologies, (DICCT)*, Dehradun, India, 2023, pp. 86–89, doi: 10.1109/DICCT56244.2023.10110252.

[13] M. Khalil, Q. Abu Al-Haija, and S. Ahmad, " Chapter 3: Healthcare IoT Networks using LPWAN", In "*Low-Power Wide Area Networks for Large Scale IoT: Architectures, Communication Protocols, and Recent Trends*," Accepted, to appear in "Advances in Cybersecurity Management" by CRC Press, Taylor, and Francis Publisher, 2023.

[14] P. K. Geo, E. S. Suraj, N. Rajendran, A. D. Kurukwar, H. Tannady and Y. Katkamwar, "IoT in Monitoring and Improving Manufacturing Process in Manufacturing Industry," *2022 4th International Conference on Inventive Research in Computing Applications (ICIRCA)*, Coimbatore, India, 2022, pp. 441–444, doi: 10.1109/ICIRCA54612.2022.9985556.

[15] Sumit and R. S. Chhillar, "IoT Enabled Security Management Approach for Intelligent Transportation System in Smart Cities," *2022 IEEE 6th Conference on Information and Communication Technology (CICT)*, Gwalior, India, 2022, pp. 1–5, doi: 10.1109/CICT56698.2022.9997813.

[16] C. Prabha and A. Pathak, "Enabling Technologies in Smart Agriculture: A Way Forward Towards Future Fields," *2023 International Conference on Advancement in Computation & Computer Technologies (InCACCT)*, Gharuan, India, 2023, pp. 821–826, doi: 10.1109/InCACCT57535.2023.10141722.

[17] T. Bhardwaj, S. Mehenge and B. S. Revathi, "Wind Turbine Power Output Forecasting Using Artificial Intelligence," *2022 International Virtual Conference on Power Engineering Computing and Control: Developments in Electric Vehicles and Energy Sector for Sustainable Future (PECCON)*, Chennai, India, 2022, pp. 1–5, doi: 10.1109/PECCON55017.2022.9851008.

[18] N. Dimitropoulos et al., "Forecasting of Short-Term PV Production in Energy Communities through Machine Learning and Deep Learning Algorithms," *2021 12th International Conference on Information, Intelligence, Systems & Applications (IISA)*, Chania Crete, Greece, 2021, pp. 1–6, doi: 10.1109/IISA52424.2021.9555544.

[19] H. Kalkha, A. Khiat, A. Bahnasse and H. Ouajji, "The Rising Trends of Smart E-Commerce Logistics," in *IEEE Access*, vol. 11, 2023, pp. 33839–33857, doi: 10.1109/ACCESS.2023.3252566.

[20] A. Miklosik, M. Kuchta, N. Evans and S. Zak, "Towards the Adoption of Machine Learning-Based Analytical Tools in Digital Marketing," in *IEEE Access*, vol. 7, 2019, pp. 85705–85718, doi: 10.1109/ACCESS.2019.2924425.

[21] M. Abdelrahim and F. Al-Turjman, "Internet of Things (IoT) and Machine Learning (ML) in Cyber-Security," *2022 International Conference on Artificial Intelligence of Things and Crowdsensing (AIoTCs)*, Nicosia, Cyprus, 2022, pp. 385–388, doi: 10.1109/AIoTCs58181.2022.00127.

[22] R. S. Subramanian, J. Aswini, S. N. Bushra, G. Nirmala and B. Maheswari, "Customer Analysis using Machine Learning with Feature Selection Approaches: A Comparative Study," *2022 International Conference on Augmented Intelligence and Sustainable Systems (ICAISS)*, Trichy, India, 2022, pp. 196–202, doi: 10.1109/ICAISS55157.2022.10011059.

[23] K. Ashok, A. B. Gurulakshmi, M. B. Prakash, R. Poornima, N. S. Sneha and V. Gowtham, "A Survey on Design and Application Approaches in Women-Safety Systems," *2022 8th International Conference on Advanced Computing and Communication Systems (ICACCS)*, Coimbatore, India, 2022, pp. 101–110, doi: 10.1109/ICACCS54159.2022.9784981.

[24] S. El-Gendy, M. S. Elsayed, A. Jurcut, .M. A. Azer, "Privacy Preservation Using Machine Learning in the Internet of Things," *Mathematics* 2023, vol. 11, 3477. https://doi.org/10.3390/math11163477

[25] N. Schizas, A. Karras, C. Karras, S. Sioutas, "TinyML for Ultra-Low Power AI and Large Scale IoT Deployments: A Systematic Review," *Future Internet* 2022, vol. 14, 363. https://doi.org/10.3390/fi14120363

[26] W. Naudé, "Artificial intelligence: neither Utopian nor apocalyptic impacts soon," *Economics of Innovation and New Technology*, 2021, vol. 30 (1), 1–23.

CHAPTER 3

Ethical Use of Machine Learning Techniques in Smart Cities

Wasswa Shafik

3.1 INTRODUCTION

The twenty-first century has witnessed unprecedented urbanization, leading to the rapid growth of smart cities (SCs) worldwide. These modern urban hubs leverage advanced technologies like the Internet of Things (IoT) and artificial intelligence (AI) to optimize infrastructure, enhance public services, and improve overall urban living. One of the key pillars underpinning the evolution of SCs is the integration of machine learning (ML) techniques [1]. ML empowers SCs to analyze vast volumes of data, make data-driven decisions, and predict future trends, fostering efficiency and sustainability across various sectors. However, this surge in ML adoption in SCs has raised ethical considerations, prompting a crucial need to examine the ethical use of ML techniques in these dynamic urban environments [2].

The SDGs, adopted by all United Nations Member States in 2015, set forth an ambitious global agenda to tackle pressing societal, economic, and environmental challenges by 2030. The SDGs encompass many objectives, including eradicating poverty, promoting sustainable cities, ensuring access to quality education and healthcare, combating climate change, and fostering social equity [3]. SCs align naturally with these goals, offering potential solutions to address complex urban issues and contribute to achieving the SDGs. However, as SCs increasingly rely on ML to drive innovation and optimize urban processes, ethical concerns have emerged, necessitating a comprehensive investigation into the responsible and equitable use of ML techniques [4].

The concept of SCs has emerged as a response to the challenges posed by rapid urbanization and the need for more efficient and sustainable urban environments. As urban populations grow, cities face numerous complex issues, such as traffic congestion, pollution, inadequate infrastructure, and resource management [5]. SCs leverage

DOI: 10.1201/9781032648323-3

innovative technologies to address these challenges, aiming to improve the quality of life for their citizens and enhance overall urban sustainability. ML techniques play a crucial role in the advancement of SCs. ML is a subset of AI that enables computer systems to learn from data and improve performance without being explicitly programmed [6].

Integrating ML techniques in SCs is driven by a compelling set of motivations that seek to address the challenges posed by rapid urbanization and pave the way for more efficient, sustainable, and livable urban environments. As urban populations grow, cities face complex issues like traffic congestion, pollution, inadequate infrastructure, and resource management [7]. ML offers a powerful solution to these problems, as its algorithms can process and analyze vast amounts of data generated by various urban systems, including traffic sensors, energy grids, public transportation, waste management, and healthcare services. This capability allows ML to derive valuable insights, identify patterns, and make predictions, enabling city administrators to instantly make data-driven decisions and optimize city operations [8]. The main components of SCs and supporting technologies are demonstrated in Figure 3.1.

One primary motivation for employing ML in SCs is to achieve greater efficiency and resource optimization. By harnessing ML analytical capabilities, cities can enhance resource allocation and usage, resulting in more efficient transportation systems, improved energy management, and streamlined public services [9]. For illustration, ML algorithms can dynamically adjust traffic signals, optimize public transportation routes, and predict energy demand patterns, reducing congestion, lowering energy consumption, and enhancing service delivery [10].

Through analyzing historical data and trends, ML can offer valuable insights into population movement, land-use patterns, and infrastructure needs. This data-driven approach empowers urban planners and policymakers to make informed decisions regarding city expansion, infrastructure development, and public service provision, contributing to

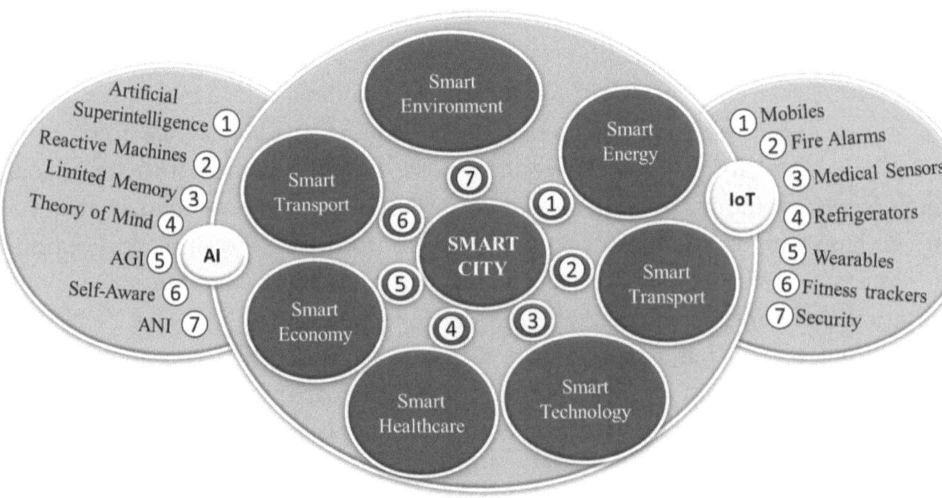

FIGURE 3.1 Main Smart City Components.

more effective and sustainable urban development [11]. Another crucial motivation for adopting ML in SCs is its capacity to improve public services. Smart applications powered by ML, such as intelligent traffic management, predictive maintenance of infrastructure, and personalized healthcare services, can significantly enhance the quality of life for city residents [12]. These applications ensure better safety, convenience, and accessibility, fostering a seamless, user-centric urban experience.

By relying on ML-generated insights, city administrators can respond promptly to evolving urban dynamics and proactively address emerging challenges. This proactive approach allows for better preparedness and quicker responses to issues like traffic incidents, natural disasters, or public health emergencies, enhancing the city's overall resilience [13]. In addition to operational advantages, ML's potential to promote sustainability and reduce the environmental impact of urban activities is a compelling incentive for SC development. ML-driven optimization of energy consumption, pollution monitoring, and climate change mitigation efforts can contribute significantly to achieving environmental sustainability goals, making cities greener and eco-friendlier [14].

3.1.1 Objectives of the Chapter

This chapter has the following objectives:

- A comprehensive exploration of the ethical considerations surrounding using ML techniques in SCs sheds light on these technologies' potential challenges and implications in urban environments.
- An in-depth analysis of the ethical dilemmas related to ML, including the examination of biases in ML algorithms, privacy concerns arising from extensive data collection, and the need for transparency and explainability in decision-making processes.
- An overview of the existing legal and regulatory frameworks governing ML applications in SCs, highlighting the importance of comprehensive guidelines and best practices to ensure responsible and ethical deployment of ML technologies.
- Case studies that illustrate the practical aspects of ethical ML implementation in specific SC applications demonstrate how proactive measures can ensure fair, transparent, and privacy-respecting use of ML in areas such as traffic management, predictive policing, and healthcare services.
- Recommendations for ethical ML implementation include establishing ethical review boards, evaluating and mitigating bias in ML algorithms, and enhancing transparency and explainability in ML models.
- Insights into the critical role of stakeholder engagement and public perception in the responsible adoption of ML in SCs, emphasizing the importance of involving communities in decision-making and building trust with citizens to foster sustainable and equitable SC development.

- Emphasis on aligning ML applications with the United Nations' SDGs, showcasing how ethical ML practices can contribute to achieving the SDGs and creating a more inclusive, resilient, and sustainable urban future.

3.1.2 Chapter Organization

The remainder of this chapter is structured in eight sections following the introduction section. Section 1.2 presents SCs and ML, entailing definitions and characteristics, an overview of ML techniques in SCs, and the potential benefits and risks of ML in SCs. Section 1.3 illustrates the ethical considerations in ML, including ethical challenges of ML in SC, fairness and bias in ML algorithms, privacy and data protection concerns, and transparency and explainability in ML models. Section 1.4 presents the legal and regulatory framework for the current laws and regulations for ML in SCs, ethical guidelines, and best practices. Moreover, the role of government and policymakers is demonstrated. Section 1.5 presents selected case studies of the ethical use of ML in SCs, such as AI-powered traffic management systems, predictive policing, crime prevention, and AI-driven healthcare services in SCs. Section 1.6 illustrates stakeholder engagement, trust, and public perception involving the community in decision-making, public awareness, and education on ethical AI, addressing concerns, and gaining public trust. Section 1.7 presents the recommendations for ethical ML implementation, including establishing ethical review boards, evaluating and mitigating bias in ML algorithms, enhancing transparency and explainability, and strengthening data privacy and security measures. Finally, Section 1.8 presents the lessons learned from the chapter and the conclusion.

3.2 SMART CITIES AND MACHINE LEARNING

SCs are urban areas that utilize advanced technologies, data, and connectivity to enhance residents' efficiency, sustainability, and quality of life. These cities leverage digital infrastructure components, including the IoT, big data (BD) analytics, AI, and ML to collect and process vast amounts of data from different sources [15]. Integrating smart technologies enables city administrators to make data-driven decisions, optimize resource allocation, improve public services, and address urban challenges effectively. The overarching goal of SCs is to create more connected, resilient, and inclusive urban environments that cater to the needs of their citizens while minimizing resource consumption and environmental impact [16].

ML is a subset of AI that focuses on developing algorithms and models capable of learning and improving from experience without being explicitly programmed. ML algorithms are designed to analyze data, identify patterns, and make predictions or decisions based on the identified patterns. This technology enables computers to automatically learn and adapt to new data, continuously improving their performance over time [17]. In SCs, ML plays a pivotal role in processing and analyzing the vast amounts of data generated by various urban systems and sensors [18]. By utilizing ML, SCs can gain valuable insights, make predictions, and optimize processes across numerous sectors, such as transportation, energy, healthcare, waste management, and public safety. ML's ability to analyze complex data sets and identify hidden patterns makes it a powerful tool for solving urban challenges and improving SC initiatives' overall efficiency and effectiveness [19].

3.2.1 Characteristics of Smart Cities

SCs exhibit vital characteristics that distinguish them from traditional urban environments. These characteristics are driven by integrating advanced technologies and data-driven approaches to address urban challenges and improve overall urban living.

SCs leverage cutting-edge technologies identical to IoT, AI, ML, BD analytics, and cloud computing. These technologies enable the collection, analysis, and utilization of vast amounts of data to make informed decisions and optimize city operations. SCs rely on data as a fundamental resource [20]. They collect and analyze data from various sources, including sensors, devices, and citizen interactions to gain insights and drive decision-making processes. Data-driven decision-making allows city administrators to respond to challenges proactively and make informed choices that improve efficiency and service delivery [21].

Sustainability is a cornerstone of SCs since they prioritize energy efficiency, renewable energy sources, and eco-friendly practices to reduce environmental impact. Smart infrastructure designs consider long-term sustainability and aim to minimize resource consumption and waste generation. Connectivity and Communication: SCs emphasize seamless connectivity among various urban systems, including transportation, energy grids, healthcare, and public safety [22]. Enhanced communication networks and data exchange facilitate real-time monitoring, coordination, and responsiveness.

Citizen-Centric Smart Cities prioritize citizen engagement and involvement in decision-making processes. Their goal is to align urban development and services with the needs and expectations of residents, ensuring that they are tailored to the preferences and requirements of the community. Mobility and Transportation: SCs focus on efficient and sustainable transportation systems [23]. They promote multimodal transportation options, integrated public transit networks, and smart traffic management to reduce congestion and improve mobility for citizens.

SCs adopt digital platforms and e-governance solutions to enhance administrative processes and service delivery. Digital governance facilitates transparent, efficient, and accessible government services for citizens [24]. Safety and Security: SCs implement advanced security measures to ensure public safety. These include using surveillance systems, emergency response technologies, and predictive policing to maintain a secure urban environment.

SCs foster innovation and support entrepreneurship. They provide opportunities for startups and businesses to develop and implement smart technologies, contributing to economic growth and job creation. Resilience and Adaptability: SCs prioritize resilience and adaptability to cope with changing urban dynamics, natural disasters, and unforeseen challenges [25]. They implement strategies to ensure continuity and rapid recovery during crises. Moreover, SCs emphasize collaboration among various stakeholders, including governments, businesses, academia, and citizens. Partnerships facilitate sharing resources, knowledge, and expertise for effective SC implementation [25].

3.2.2 Potential Benefits and Risks of Machine Learning in Smart Cities

ML offers various potential benefits in SCs, enabling data-driven decision-making and optimization of urban systems. However, its implementation also has inherent risks that must be carefully addressed.

One of the primary benefits of ML in SCs is its ability to process and analyze vast amounts of data from various sources, including sensors, social media, and public records. This data-driven approach empowers city administrators to make informed decisions based on real-time insights and trends, leading to more efficient and effective urban management [26]. ML-powered applications can significantly enhance public services in SCs. For instance, ML can optimize traffic management by predicting congestion patterns and adjusting traffic signals accordingly [27]. Similarly, ML algorithms can facilitate predictive maintenance of urban infrastructure, reducing downtime and ensuring reliable services for residents.

ML is crucial in optimizing resource consumption and waste management in SCs. By analyzing energy usage, water consumption, and waste generation data, ML algorithms can identify patterns and predict demand, enabling cities to allocate resources more efficiently and reduce their environmental footprint [28]. Also, ML can bolster safety and security in SCs by analyzing data from surveillance cameras, social media, and other sources to detect anomalies and potential security threats. This proactive approach enables authorities to respond swiftly to incidents and prevent crime, ultimately creating a safer urban environment [29]. The role and benefits of different SC domains are presented in Figure 3.2.

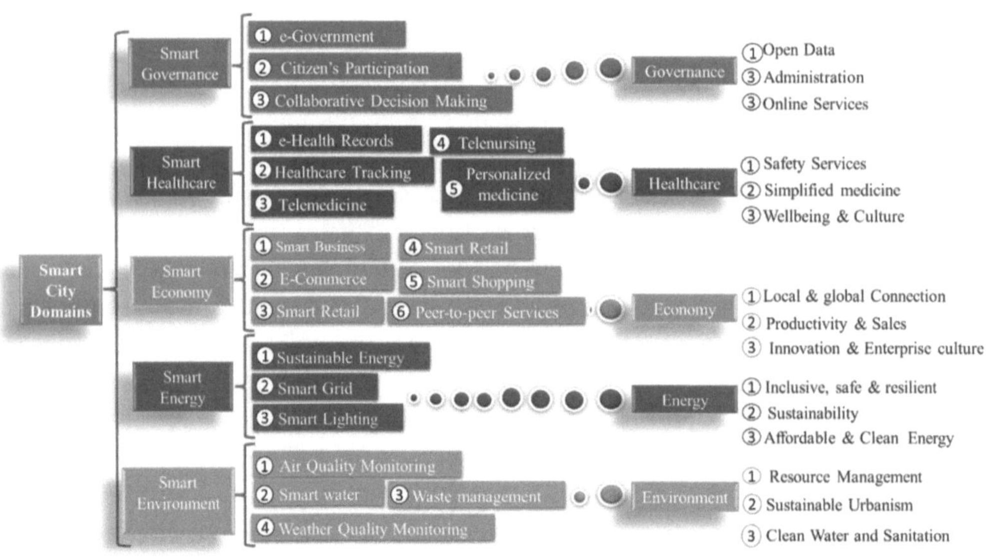

FIGURE 3.2 Role of Technology in the Smart Cities.

ML algorithms optimize transportation systems by predicting travel demand, analyzing traffic patterns, and managing public transit schedules. This results in reduced travel time, improved traffic flow, and better mobility options for residents, contributing to a more seamless and efficient transportation network, and ML's real-time monitoring capabilities are instrumental in addressing urban challenges promptly [30]. For instance, ML can detect air quality issues or traffic accidents in real-time, allowing for swift response and mitigation measures.

The extensive data collection in SCs raises concerns about data privacy and security. Improper handling of sensitive data could lead to privacy breaches and expose citizens to potential identity theft or surveillance risks. ML algorithms learn from historical data, and if that data contains biases, the algorithms may perpetuate and amplify those biases in decision-making processes [31]. This can lead to unfair or discriminatory outcomes, impacting certain communities disproportionately.

ML, bottomless learning models, are highly complex and challenging to interpret. The lack of transparency and explainability in ML decisions can undermine public trust and accountability. Overreliance on ML and technology may lead to an overemphasis on data-driven solutions, potentially overlooking the importance of human judgment, social considerations, and community engagement in urban planning and decision-making [32]. Implementing ML in a complex urban environment requires seamless integration with existing systems and interoperability among various data sources. Integration challenges can hinder the effectiveness and scalability of ML applications in SCs [33]. While ML can bring numerous benefits to SCs, there is a risk of exacerbating economic disparities if specific segments of the population do not have access to technology or do not benefit from the improvements brought about by ML-driven initiatives [34].

It is essential to adopt a proactive and ethical approach to fully harness the potential benefits of ML in SCs while mitigating its associated risks. This includes transparent and accountable data governance, continuous monitoring and evaluation of ML models, efforts to address biases, and public engagement to ensure that ML technologies align with the needs and values of the communities they serve [35]. By carefully navigating these considerations, SCs can leverage ML's power to create more inclusive, sustainable, and equitable urban environments.

3.3 ETHICAL CONSIDERATIONS IN MACHINE LEARNING

This section presents the most outstanding ethical challenges of ML in SCs. Ethical considerations in ML are paramount, especially in SCs, where ML is increasingly deployed to make critical decisions impacting people's lives and shaping urban environments.

3.3.1 Fairness and Bias in Machine Learning Algorithms

Ethical considerations in ML address the issue of bias and fairness, ensuring that ML algorithms do not perpetuate or amplify existing societal biases present in the training data. By actively mitigating bias, SCs can make fair and equitable decisions for all residents, regardless of race, gender, or other personal attributes [36]. This leads to a more inclusive urban environment where all citizens receive equal opportunities and treatment in accessing public services and resources.

3.3.2 Privacy and Data Protection Concerns

Ethical ML implementation in SCs prioritizes protecting citizens' privacy and personal data. Stricter data governance and privacy policies safeguard individuals' sensitive information from unauthorized access or misuse. This, in turn, fosters trust between the city administration and its residents, encouraging greater participation and engagement in SC initiatives [37].

3.3.3 Transparent Decision-Making Processes

Ethical ML deployment emphasizes transparency and explainability in the decision-making processes of ML algorithms. Transparent models provide clear reasons for their predictions, enabling city administrators and citizens to understand how decisions are made [38]. This transparency enhances accountability, as decisions can be scrutinized and validated, instilling confidence in the fairness and reliability of ML-driven urban management.

3.3.4 Informed Consent and Citizen Autonomy

Addressing ethical considerations ensures citizens' consent is obtained before using their data in ML applications. Providing individuals the autonomy to control their data empowers them to make informed choices about data sharing and usage [39]. Respecting individuals' preferences and privacy rights fosters a positive relationship between the city and its citizens, reinforcing citizen engagement and cooperation in building the SC.

3.3.5 Responsible and Accountable Governance

Ethical ML practices establish clear lines of accountability for the outcomes of ML algorithms. Responsible governance mechanisms are implemented to monitor and assess ML applications' impact on the city and its residents. This accountability holds stakeholders, including government agencies and private entities, responsible for the ethical implications of ML deployment, leading to the more responsible and ethical use of ML technologies in SCs [40].

3.3.6 Enhanced Safety and Security

Ethical ML models enhance public safety and security in SCs. ML-driven surveillance systems can detect anomalies and potential security threats, aiding law enforcement agencies in crime prevention and emergency response [41]. Responsible ML implementation ensures that these technologies are used in ways that respect individual privacy and civil liberties while protecting public safety.

3.3.7 Inclusive and Socially Beneficial Technology

Considering the social impact and inclusivity of ML technologies ensures that SC initiatives cater to the needs of all citizens, including marginalized communities [42]. Ethical ML implementation promotes social equality and reduces disparities, ensuring that all, regardless of socioeconomic status, share the benefits of SC technologies.

3.3.8 Environmental Sustainability

Ethical ML practices consider the environmental impact of ML applications. By promoting energy-efficient practices, reducing model complexity, and optimizing data centers, SCs can minimize the carbon footprint of ML technologies [43]. This aligns with broader sustainability goals, contributing to a greener, eco-friendlier urban environment.

3.4 LEGAL AND REGULATORY FRAMEWORK

The legal and regulatory framework, ethical guidelines and best practices, and the role of government and policymakers on ML in SCs are crucial to governing the responsible and ethical deployment of ML technologies. As ML becomes increasingly integrated into urban systems, it is essential to establish guidelines, laws, and standards to address potential risks and ensure compliance with ethical principles.

3.4.1 Current Laws and Regulations for Machine Learning in Smart Cities

The legal and regulatory framework for ML in SCs typically encompasses the following aspects.

Data protection laws play a fundamental role in governing the collecting, storing, and processing personal data in SCs. Regulations such as those in Europe and similar laws in other regions require organizations to obtain explicit consent for data collection, ensure data accuracy, and provide individuals with the right to access and control their data [44]. Security and Cybersecurity Regulations: ML models and data repositories in SCs need robust security measures to protect against cyber threats and data breaches. Legal frameworks encompassing cybersecurity regulations dictate the requirements for securing data, systems, and networks to safeguard sensitive information and prevent unauthorized access [45].

Some jurisdictions may require transparency and explainability standards for ML algorithms used in critical decision-making processes. These regulations ensure that ML models' decision-making processes are interpretable and can be understood by experts and the public. Some countries or organizations may develop ethical guidelines and standards specific to ML in SCs [46]. These guidelines address fairness, bias mitigation, and accountability in developing and deploying ML technologies in urban environments. Human Rights and Civil Liberties Protections: ML algorithms should not infringe on human rights and civil liberties. Legal frameworks may include provisions that protect individuals from discriminatory practices, unauthorized surveillance, or any ML-driven actions that violate fundamental human rights [47].

Legal frameworks may define governance structures and accountability mechanisms for ML applications in SCs. This could involve establishing regulatory bodies or ethical review boards to oversee ML deployments, assess risks, and ensure compliance with ethical standards. Intellectual Property and Data Ownership: Legal frameworks may address issues related to intellectual property rights for ML algorithms and data ownership [48]. Regulations could determine ownership rights for data generated by public infrastructure and govern intellectual property claims on ML models developed through public funding. Liability and Responsibility: Legal frameworks may clarify liability and responsibility for ML outcomes in SCs. Determining who is accountable for errors or unintended consequences of ML algorithms is crucial to ensure appropriate recourse and responsibility [49].

For SCs operating across national borders, legal frameworks may address data transfer across jurisdictions while ensuring compliance with data protection laws in different regions. Public Engagement and Transparency Requirements: Some legal frameworks may require public engagement and transparency in deploying ML technologies in SCs. This could involve public consultations, citizen participation, and disclosing ML applications' intentions and potential impacts [50]. As ML technologies in SCs continue to evolve, legal and regulatory frameworks must adapt to address emerging challenges and ethical considerations. A comprehensive and well-defined legal framework fosters responsible innovation, protects citizens' rights, and ensures that ML contributes positively to developing inclusive, sustainable, and trustworthy SCs [50].

3.4.2 Ethical Guidelines and Best Practices

These guidelines provide a framework for developers, city administrators, and stakeholders to adhere to ethical principles, mitigate potential risks, and maximize the positive impact of ML in urban environments. ML algorithms should be designed and evaluated to minimize biases and ensure fairness in decision-making. Rigorous testing and auditing of ML models are essential to identify and address any biases that may arise from the training data [51]. Techniques such as data preprocessing, fairness-aware algorithms, and bias-aware evaluation can help promote fairness and reduce discrimination.

Transparency and Explainability where ML models are deployed in SCs should be transparent and explainable. Providing clear explanations for ML predictions and decisions helps build trust with the public and stakeholders. Adopting interpretable ML algorithms and providing understandable explanations can enhance transparency and accountability. Respecting data privacy is paramount in SCs [52]. Adhering to data protection regulations, obtaining informed consent from individuals for data usage, and implementing robust data security measures are crucial to safeguarding citizens' privacy rights.

Establishing clear lines of accountability and governance for ML applications is essential. This may involve creating oversight committees, ethical review boards, or regulatory bodies responsible for monitoring and evaluating ML deployments to ensure adherence to ethical guidelines. Involving citizens and communities in the decision-making process for ML deployments fosters inclusive SC development [53]. Public consultations, citizen engagement, and transparency in decision-making build trust and ensure that ML technologies align with societal values and needs.

Continuous Monitoring and Evaluation, in terms of regularly monitoring and evaluating ML models in real-world settings, is crucial for identifying and addressing ethical issues that may emerge over time. Continuous evaluation allows for prompt adjustments and improvements to maintain ethical standards. SCs should collect data responsibly, ensuring that data collected is relevant, necessary, and obtained through ethical means [54]. Data should be used solely for intended purposes and not repurposed without proper consent. Promoting ML literacy among citizens helps them understand the capabilities and limitations of ML technologies. Educating the public about how ML is used in SCs encourages responsible usage, and fosters informed participation in decision-making processes [55].

Prioritizing energy-efficient ML algorithms and infrastructure helps mitigate the environmental impact of ML applications in SCs. Collaborating with diverse stakeholders, including academia, industry, and non-profit organizations, promotes ethical discussions, knowledge sharing, and collaborative problem-solving in deploying ML technologies [56]. Implementing ethical guidelines and best practices ensures that ML technologies in SCs are developed and utilized in a manner that aligns with human values, protects individual rights, and contributes positively to the well-being of urban communities. Responsible ML deployment fosters trust, resilience, and sustainability in SC development, creating more inclusive, equitable, and prosperous urban environments [57].

3.4.3 The Role of Government and Policymakers

The role of government and policymakers is crucial in the ethical and responsible deployment of ML technologies in SCs. As ML continues to play an increasingly significant role in urban development, policymakers have the responsibility to create a conducive regulatory environment that maximizes the benefits of ML while addressing potential risks and ethical considerations [24]. The following are some critical roles of government and policymakers in the context of ML in SCs:

Government and policymakers play a central role in developing ethical guidelines and regulations for using ML in SCs. These guidelines should address bias mitigation, fairness, transparency, data privacy, and accountability. By setting clear ethical standards, governments provide a framework that encourages responsible and ethical ML practices. Governments are responsible for creating legal and regulatory frameworks that govern data collection, storage, and use in SCs [25]. This includes data protection laws, cybersecurity regulations, and guidelines for cross-border data transfer, ensuring that citizens' privacy rights are protected.

Government agencies are tasked with monitoring and enforcing compliance with ethical guidelines and regulations. This involves assessing ML applications for adherence to ethical standards, investigating complaints, and imposing penalties for non-compliance. Policymakers can support research and innovation in ML technologies through funding, grants, and collaborations with academia and industry [26]. Encouraging research in ethical AI and ML helps develop cutting-edge solutions that align with SC's goals and values.

Moreover, policymakers are vital in promoting public awareness and education about ML technologies. This involves informing citizens about the benefits and risks of ML, their rights regarding data privacy, and the ethical considerations associated with ML in SCs [27]. Governments can foster collaboration and partnerships among stakeholders, including city authorities, businesses, researchers, and citizens. Collaborative efforts can lead to more holistic and responsible ML implementations in SC projects.

Government agencies can conduct impact assessments of ML technologies deployed in SCs. These assessments evaluate ML applications' social, economic, and environmental impact, helping identify potential risks and benefits [28]. Policymakers can promote open data initiatives, facilitating the sharing of non-sensitive, anonymized data for research and innovation. Open data can contribute to developing ML models and promote transparency in decision-making processes [29]. Governments need to address concerns related to the impact of ML on employment and the economy. Implementing policies for retraining and upskilling workers affected by automation and supporting businesses transitioning to AI-based technologies can foster inclusive economic growth.

Since ML technologies cross borders, governments can collaborate internationally to harmonize ethical standards and regulatory approaches. International cooperation ensures consistent ethical practices in ML deployment across various SCs globally. By assuming these roles, the government and policymakers can create an environment that fosters ethical and responsible ML deployment in SCs [30]. Their active involvement in developing comprehensive policies and guidelines ensures that ML technologies contribute positively to building sustainable, inclusive, and people-centric SCs.

3.5 CASE STUDIES: ETHICAL USE OF MACHINE LEARNING IN SMART CITIES

This section demonstrates some of the three most famous case study applications of ML in the SCs.

3.5.1 Case Study 1: AI-Powered Traffic Management System

This case study will explore implementing an AI-powered traffic management system in a SC. The SC aims to leverage ML technologies to optimize traffic flow, reduce congestion, and enhance overall mobility for its residents [38]. The AI-powered traffic management system is part of the city's broader initiative to create a sustainable, efficient, and people-centric urban environment, as illustrated in Figure 3.3.

3.5.1.1 Objectives of Case Study 1

The primary objectives of implementing the AI-powered traffic management system are as follows:

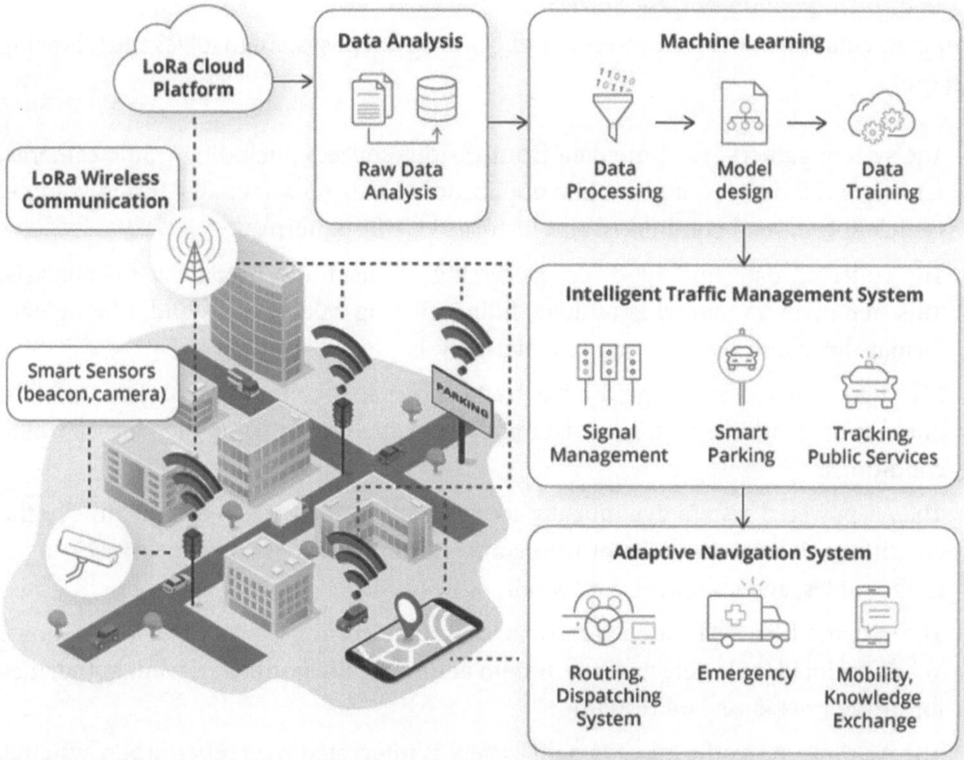

FIGURE 3.3 AI Traffic Management System using LoRa Cloud Platform.

Source: Intellias.

- The system aims to optimize traffic flow by analyzing real-time traffic data, identifying congestion points, and adjusting traffic signals and flow patterns accordingly [27]. The objective is to reduce traffic bottlenecks and improve the overall efficiency of the city's transportation network.
- The AI-powered system aims to mitigate congestion on major roadways and intersections by predicting traffic patterns and anticipating peak hours. This will result in shorter travel times, decreased fuel consumption, and lower greenhouse gas emissions.
- The system incorporates AI-based video analytics to simultaneously detect and respond to traffic violations and accidents. By alerting traffic authorities promptly, the system aims to enhance road safety and reduce the number of accidents in the city.
- The AI-powered traffic management system relies on data-driven decision-making processes. It analyzes vast amounts of traffic data, weather conditions, and historical patterns to make informed and optimized traffic management decisions.

3.5.1.2 Implementation of Case Study 1

The implementation of the AI-powered traffic management system involves the following key steps:

- The system gathers real-time data from various sources, including traffic cameras, sensors, GPS devices, and mobile applications. This data includes traffic volume, vehicle speed, road conditions, and historical traffic patterns.
- The collected data undergoes preprocessing to clean and prepare it for analysis. This step involves removing outliers, filling missing values, and standardizing data formats to ensure accurate and reliable analysis.
- ML algorithms develop predictive traffic flow and congestion pattern models. These models utilize historical data and real-time inputs to predict future traffic conditions.
- The AI-powered system dynamically adjusts traffic signals based on real-time traffic conditions. It optimizes signal timings to minimize congestion, prioritizes high-traffic routes, and enhances the overall traffic flow.
- The system utilizes AI-based video analytics to detect accidents or traffic violations. When an incident is detected, the system automatically notifies relevant authorities for prompt response and resolution.
- The AI-powered traffic management system is integrated with other urban systems, such as public transportation and emergency services. This integration enables coordinated responses and enhances overall urban efficiency.
- The system is continuously monitored to assess its performance and effectiveness. Regular evaluations help identify areas for improvement and potential adjustments to optimize the system further.

3.5.1.3 Benefits of Case Study 1

The implementation of the AI-powered traffic management system brings several benefits to the SC:

- By optimizing traffic flow and signal timings, the system significantly reduces congestion, leading to shorter travel times and improved traffic efficiency.
- Real-time incident detection and response contribute to improved road safety and reduced accident rates in the city.
- Reduced congestion and optimized traffic flow result in lower fuel consumption and greenhouse gas emissions, promoting environmental sustainability.
- The traffic management system provides valuable data and insights to inform urban planning decisions, such as identifying areas for infrastructure improvements and future transportation developments.
- The system's ability to reduce traffic congestion and enhance safety contributes to residents' better quality of life, making the city more attractive and livable.

3.5.1.4 Lessons Learned from Case Study 1

Implementing an AI-powered traffic management system demonstrates the potential of ML technologies in creating smart and efficient urban environments [42]. By optimizing traffic flow, reducing congestion, enhancing safety, and employing data-driven decision-making processes, the SC significantly improves its transportation network's overall performance and benefits its residents and visitors. This case study is a prime example of how AI and ML can be harnessed to address urban challenges and move towards a more sustainable and people-centric SC future.

3.5.2 Case Study 2: Predictive Policing and Crime Prevention

This case study focuses on implementing predictive policing and crime prevention using machine ML technologies in a SC. The SC uses data-driven approaches to enhance public safety, reduce crime rates, and optimize law enforcement resources [9]. The city aims to proactively identify potential crime hotspots and allocate resources more effectively to prevent criminal activities by deploying ML algorithms to analyze historical crime data and real-time information.

3.5.2.1 Objectives of Case Study 2

The primary objectives of implementing predictive policing and crime prevention are as follows:

- The ML-based system aims to predict crime incidents by analyzing historical crime data, identifying patterns, and forecasting potential hotspots. Predictive models are developed to anticipate when and where crimes will likely occur, enabling law enforcement to take preemptive action.
- Through predicting crime hotspots, law enforcement agencies can optimize the allocation of personnel and resources. Focused deployment of officers to high-risk areas can deter criminal activities and enhance overall law enforcement efficiency.
- The predictive policing system enables law enforcement to adopt a proactive approach by addressing potential criminal activities before they escalate. Early intervention and targeted preventive measures contribute to a safer urban environment.
- The SC fosters community engagement by involving residents in crime prevention efforts [37]. Citizens can access crime-related information and safety tips, contributing to a collaborative and vigilant community.

3.5.2.2 Implementation of Case Study 2

The implementation of predictive policing and crime prevention involves the following key steps:

- The system collects and integrates data from various sources, including historical crime records, emergency calls, surveillance cameras, social media, and other relevant datasets. This data is used to train and update the ML models.

- Collected data undergoes preprocessing to ensure its quality, consistency, and compatibility. Data cleaning, normalization, and feature engineering are performed to prepare the data for ML analysis.
- ML algorithms, such as supervised learning, clustering, and anomaly detection, are employed to develop predictive crime models. These models learn from historical crime data to identify patterns and predict future crime.
- The predictive models generate crime hotspot maps that indicate areas with higher likelihoods of criminal activities. These maps are continuously updated based on real-time data, enabling dynamic allocation of resources.
- Resource Allocation: Law enforcement agencies use hotspot maps to allocate police officers and resources to high-risk areas strategically. This targeted approach optimizes law enforcement efforts and response times.
- The SC engages with the community through various channels, providing safety tips, reporting mechanisms, and crime awareness programs. Encouraging citizens to participate actively in crime prevention strengthens community bonds and improves safety.
- The system is continually evaluated to assess its performance and effectiveness. Regular feedback from law enforcement and the community helps refine the predictive models and enhance crime prevention strategies.

3.5.2.3 Benefits of Case Study 2

The implementation of predictive policing and crime prevention brings several benefits to the SC:

- Proactive crime prevention strategies and resource optimization reduce crime rates, making the city safer for its residents and visitors.
- Predictive policing allows law enforcement to use resources more efficiently by directing efforts to areas with higher crime probabilities, leading to better crime resolution rates.
- By involving the community in crime prevention efforts, the SC fosters a sense of security and empowerment among its residents.
- Optimized resource allocation avoids unnecessary expenditures and maximizes the impact of law enforcement efforts, leading to potential cost savings for the city.
- The predictive policing system facilitates data-driven decision-making processes, enabling law enforcement agencies to make informed choices based on real-time and historical crime data.

3.5.2.4 Lessons Learned from Case Study 2

Implementing predictive policing and crime prevention demonstrates the potential of ML technologies in creating safer and more secure SCs. By leveraging data analytics and predictive models, the city can proactively address criminal activities, allocate law enforcement resources efficiently, and engage the community in crime prevention efforts [47]. This case study exemplifies how ML can improve public safety, promote community collaboration, and enhance the overall quality of life in the SC.

3.5.3 Case Study 3: AI-Driven Healthcare Services in Smart Cities

This case study focuses on the implementation of AI-driven healthcare services in a SC. The SC aims to leverage AI technologies to revolutionize healthcare delivery, enhance patient outcomes, and optimize healthcare resources as illustrated in Figure 3.4. By harnessing AI algorithms and data analytics, the city aims to provide personalized,

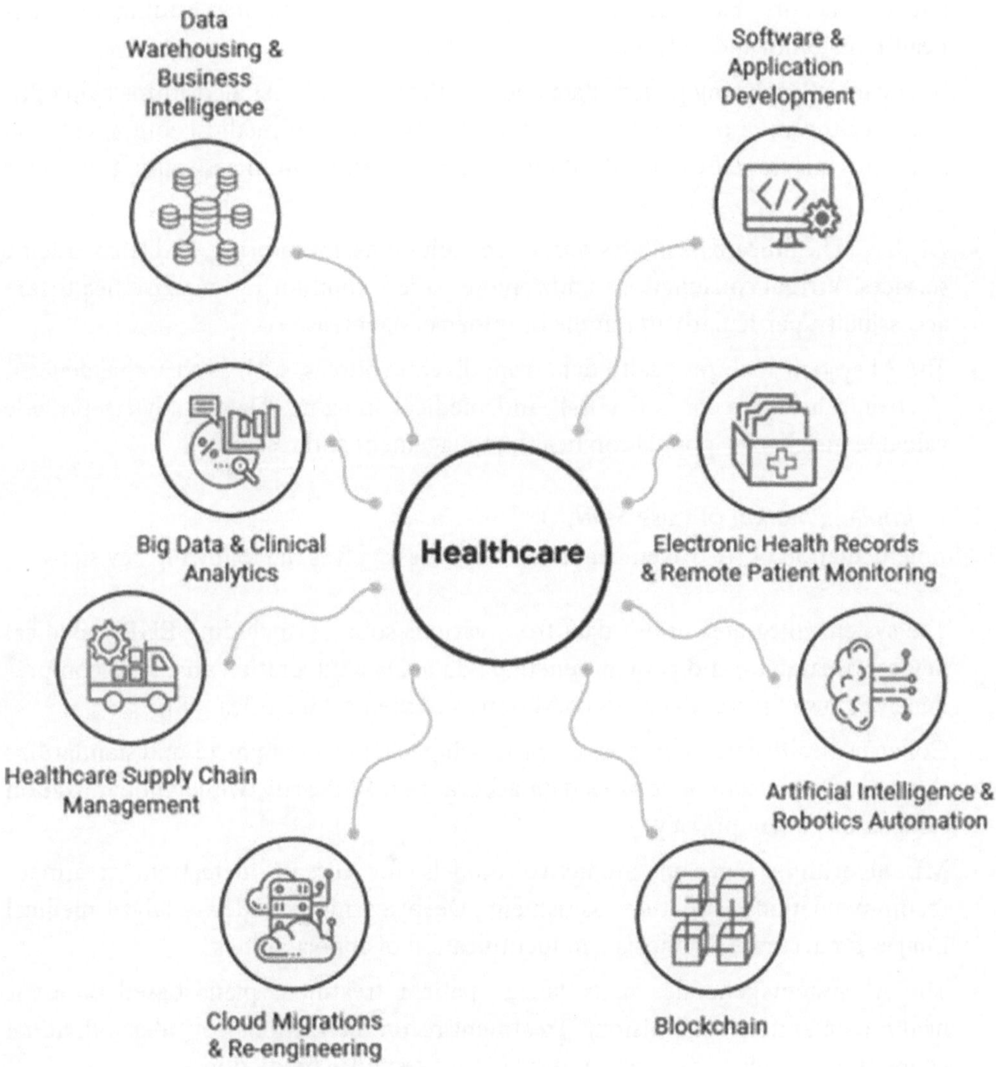

FIGURE 3.4 AI-Driven Healthcare Services Industry.

Source: kcsitglobal.

efficient, and accessible healthcare services to its residents, fostering a healthier and more resilient urban community [49].

3.5.3.1 Objectives of Case Study 3

The primary objectives of implementing AI-driven healthcare services are as follows:

- The AI-driven healthcare system aims to personalize medical treatments and interventions based on individual patient data, medical history, and genetic information [26]. Personalized healthcare ensures patients receive tailored treatments most effective for their specific health conditions.
- The AI system analyzes vast amounts of health data to detect early signs of diseases and risk factors. Early detection enables timely interventions, leading to better health outcomes and reduced healthcare costs.
- By means of analyzing patient data and healthcare trends, AI algorithms help optimize healthcare resource allocation, such as hospital beds, medical equipment, and staff. This ensures efficient utilization of resources and minimizes waiting times for patients.
- AI-driven healthcare facilitates remote consultations, monitoring, and telemedicine services. Virtual consultations and remote patient monitoring improve healthcare accessibility, particularly in remote or underserved areas.
- The AI system analyzes health data from diverse sources, such as wearable devices, electronic health records (EHRs), and medical imaging. Data analytics provide valuable insights for population health management and research.

3.5.3.2 Implementation of Case Study 3

The implementation of AI-driven healthcare services involves the following key steps:

- The system integrates health data from various sources, including EHRs, medical devices, wearables, and patient-generated data. Data integration ensures a comprehensive view of patient's health conditions and medical histories [38].
- Collected health data undergoes preprocessing to clean, anonymize, and standardize the data. Data cleaning ensures data accuracy and integrity, while anonymization safeguards patient privacy.
- ML algorithms develop predictive models for disease detection, treatment recommendations, and risk assessment. Deep learning models analyze medical images for accurate diagnosis and identification of abnormalities.
- The AI system generates personalized patient treatment plans based on their health data and medical history. Treatment recommendations consider individual characteristics, genetic factors, and treatment response prediction.
- The AI system establishes early warning systems for disease detection and prevention. The system can identify potential health risks by analyzing patient data at present and alert healthcare providers for timely interventions.

- The AI-driven healthcare system integrates telemedicine platforms for remote consultations and monitoring. Patients can access healthcare services from home, reducing the need for in-person visits and enhancing healthcare accessibility.
- The system analyzes population health data to identify health trends, prevalent diseases, and risk factors in the community. Population health insights help city authorities make informed decisions for public health initiatives.
- The AI-driven healthcare system continuously learns from new data and feedback to refine its models and recommendations. This continuous learning process ensures that the system remains up-to-date and effective in delivering healthcare services.

3.5.3.3 Benefits of Case Study 3

The implementation of AI-driven healthcare services brings several benefits to the SC:

- Personalized treatment plans and early disease detection improve patient outcomes and overall health [16].
- Telemedicine services and remote consultations enhance healthcare access, particularly for residents in remote or underserved areas, promoting healthcare equity.
- Optimized resource allocation and predictive analytics help healthcare facilities operate more efficiently and reduce patient waiting times.
- Early disease detection and prevention can save healthcare expenditures by avoiding expensive treatments for advanced diseases.
- Population health insights aid city authorities in making data-driven decisions for public health initiatives and resource planning.

3.5.3.4 Lessons Learned from Case Study 3

Implementing AI-driven healthcare services showcases the potential of AI in revolutionizing healthcare delivery in SCs. By leveraging data analytics, personalized treatment plans, and telemedicine integration, the city can provide efficient, accessible, and patient-centric healthcare services [26]. This case study demonstrates how AI technologies improve patient outcomes, enhance healthcare access, and optimize resource utilization, making the SC a healthier and more resilient urban community.

3.6 STAKEHOLDER ENGAGEMENT AND PUBLIC PERCEPTION

Stakeholder engagement and public perception are pivotal in successfully implementing AI-driven SC initiatives. As AI technologies are integrated into various aspects of urban life, involving diverse stakeholders, and addressing public perceptions is essential to foster trust, transparency, and acceptance of AI applications [50]. This case study focuses on how stakeholder engagement and public perception can impact AI-driven SC projects and explores strategies to ensure positive public reception and collaboration among stakeholders [38]. Involving various stakeholders, including government officials, citizens, industry representatives, academia, and community organizations, in decision-making ensures diverse perspectives and a holistic approach to SC development [51].

Engaging private sector partners and technology companies foster collaboration in developing and deploying AI-driven solutions. Public-private partnerships can leverage the expertise and resources of both sectors for more effective SC implementations [18]. Conducting community outreach programs and public consultations allows citizens to voice their concerns, expectations, and needs related to AI technologies in SCs. Listening to community feedback enables project customization to meet local requirements. Ensuring transparent communication about AI technologies, their applications, and potential impacts helps build trust with stakeholders [9]. Clear communication about data usage, privacy protection, and ethical considerations is crucial for public acceptance.

Raising public awareness and educating citizens about AI technologies is vital to demystify misconceptions and enhance understanding. Educational campaigns can highlight AI's potential benefits, dispel myths, and address privacy and job displacement concerns. Implementing pilot projects and showcasing tangible benefits of AI-driven solutions in SCs helps improve public perception [27]. Demonstrating successful use cases builds confidence in AI technologies and their positive impact on urban living. Developing and adhering to an ethical AI framework emphasizes responsible AI deployment and addresses public apprehensions about biased or discriminatory AI applications. Ethical guidelines demonstrate a commitment to align AI technologies with societal values. Also, implementing robust data protection measures and ensuring stringent privacy policies alleviate public concerns about data misuse or surveillance [28].

Impact, effective stakeholder engagement, and transparent communication lead to a positive public perception of AI-driven SC projects. Citizens feel more included, informed, and confident in the benefits and ethics of AI technologies. Building trust among stakeholders fosters greater acceptance of AI applications in SCs [16]. When citizens and communities have confidence in the technologies, they are more likely to participate actively in SC initiatives. Positive public perception and stakeholder collaboration contribute to the long-term sustainability of AI-driven SC projects. Engaging stakeholders in decision-making helps identify sustainable solutions that meet the city's evolving needs. Addressing public concerns and engaging diverse stakeholders ensure that AI-driven SC projects prioritize inclusivity and equity [35]. This leads to solutions that cater to the needs of all residents and communities, promoting a more inclusive urban environment.

3.7 RECOMMENDATIONS FOR ETHICAL MACHINE LEARNING IMPLEMENTATION

We recommend the following ethical machine learning implementations.

3.7.1 Establishing Ethical Review Boards

To ensure ethical ML implementation in SCs, it is essential to establish independent Ethical Review Boards (ERBs) composed of experts in AI ethics and data privacy and representatives from various stakeholder groups, including the public [39]. ERBs will review AI projects, assess their ethical implications, and provide guidance on potential risks and benefits. The involvement of ERBs fosters accountability, transparency, and adherence to ethical standards in developing and deploying ML technologies.

3.7.2 Evaluating and Mitigating Bias in Machine Learning Algorithms

Bias in ML algorithms can perpetuate discrimination and unfair treatment. Therefore, it is crucial to conduct thorough evaluations of ML models to identify and mitigate biases. Data for train algorithms should be diverse, representative, and free of discriminatory attributes [17]. Regular audits of the algorithms' performance should be carried out to ensure fairness and to rectify biases if detected. Implementing bias-aware algorithms and conducting continuous fairness assessments will help build equitable AI systems.

3.7.3 Enhancing Explainability and Transparency

AI-driven SC applications should prioritize transparency and explainability to gain public trust and ensure accountability. ML models should be designed with interpretability, enabling stakeholders to understand how decisions are made [38]. Providing clear explanations for the reasoning behind AI-generated outcomes allows citizens to assess and validate the fairness and validity of AI technologies. Transparent AI systems build confidence and facilitate better collaboration between the city and its residents.

3.7.4 Strengthening Data Privacy and Security Measures

Protecting citizens' data privacy is paramount in ethical ML implementation. SCs must adhere to stringent data privacy regulations, ensuring data is collected, stored, and processed securely. Strong encryption, data anonymization, and access controls will safeguard sensitive information [23]. Furthermore, data sharing and access policies should be clearly defined to prevent unauthorized use. SCs must prioritize data protection to instil public trust in the responsible use of AI-driven technologies. Ethical implementation of ML technologies in SCs is essential to build sustainable, inclusive, and trustworthy urban environments [49]. By establishing ERBs, addressing bias, promoting transparency, and prioritizing data privacy, SCs can create a foundation for responsible AI adoption. These recommendations enable ethical decision-making, foster public trust, and empower SCs to harness the full potential of ML in advancing urban living.

3.8 CONCLUSION

The studies on ethical use of ML techniques in SCs provides valuable insights and lessons learned. Ethical considerations, stakeholder engagement, transparency, bias mitigation, data privacy, continuous monitoring, and public perception are integral to the responsible implementation of AI technologies.

3.8.1 Lessons Learned from the Chapter

By incorporating the following lessons into SC development, urban environments can leverage AI's transformative potential to create more inclusive, equitable, and sustainable cities for the future.

- Case study 1 highlights the criticality of ethical considerations in deploying ML techniques in SCs. Ethical guidelines and best practices are essential to ensuring that AI technologies align with societal values, protect individual rights, and contribute positively to urban development.
- The study underscores the significance of engaging diverse stakeholders, including citizens, government officials, industry representatives, and researchers [6][10][21–32]. Collaborative decision-making fosters inclusivity, enables a holistic approach to SC development [41–44], and ensures that AI solutions meet the needs and expectations of all stakeholders [35–39].
- The study emphasizes the importance of transparency and explainability in AI-driven SC applications. Transparent AI models, with clear explanations of their decision-making processes, build trust with the public and enhance accountability in using AI technologies.
- The study highlights the need to evaluate and mitigate biases in ML algorithms. Implementing bias-aware algorithms and using diverse and representative datasets helps minimize discriminatory outcomes and ensures fairness in AI applications.
- The study recognizes the significance of robust data privacy and security measures in protecting citizens' personal information. SCs must prioritize data protection to build public confidence in the responsible use of AI technologies and maintain the privacy rights of individuals.
- The study emphasizes the importance of continuously monitoring and evaluating AI systems to identify and address ethical issues as they arise. Regular assessments enable SCs to make necessary adjustments, refine AI models, and improve the overall performance of ML technologies.
- The study highlights the influence of public perception on the success of AI-driven SC initiatives. Raising public awareness about AI technologies, addressing misconceptions, and demonstrating tangible benefits are crucial to gaining public acceptance and support for AI deployments.

3.8.2 Future Trends and Final Thoughts

The SDGs present a transformative agenda for creating a more sustainable and equitable world by 2030, and the emergence of SCs has utilized cutting-edge technologies like ML to enhance urban living [52]. However, the increasing reliance on ML techniques in SC applications raises ethical concerns that must be addressed to ensure responsible and equitable implementation. This chapter thoroughly examined the ethical use of ML techniques in SCs, highlighting the need for proactive measures to tackle potential challenges. The chapter provided essential context, exploring the exponential growth of SCs and the pivotal role of ML in driving transformative changes across various sectors [53]. It delineated the scope of research, focusing on the ethical dimension of ML applications in SC environments. The chapter elaborated on foundational concepts of SCs and ML, defining the characteristics and the diverse applications of ML, including predictive analytics and optimization algorithms. The potential benefits and risks of integrating ML in SCs were critically analyzed, emphasizing the ethical implications in shaping urban experiences.

Ethical considerations formed a central theme, addressing multifaceted challenges such as biases in ML algorithms, privacy concerns related to extensive data collection, and transparency and explainability in decision-making processes [54]. The exploration of ethical dilemmas deepened the understanding of integrating ML in a manner aligned with societal values and principles. The chapter also delved into the existing legal and regulatory frameworks governing ML applications, emphasizing the necessity of comprehensive guidelines and best practices for responsible deployment [55]. Case studies of specific applications in SCs were presented, examining an AI-powered traffic management system, predictive policing and crime prevention, and AI-driven healthcare services.

3.8.3 Conclusion

The analysis of each case study provided valuable insights into ethical considerations, showcasing proactive measures to ensure fair, transparent, and privacy-respecting ML applications. The pivotal role of stakeholder engagement and public perception in ethical ML adoption was discussed, emphasizing the importance of involving the community in decision-making processes, raising public awareness of ethical AI usage, and building trust with citizens to foster responsible and sustainable SC developments. In conclusion, the study highlighted the criticality of ethical considerations in developing SCs driven by ML technologies. By integrating ethical guidelines, transparent decision-making, and stakeholder collaboration, SCs can navigate potential challenges and harness the full potential of ML for creating inclusive, sustainable, and people-centric urban environments. Emphasizing ethical principles ensures that SCs align with societal values, protect individual rights, and advance toward a more equitable and prosperous future.

REFERENCES

[1] Sadoughi F, Kazemy Z, Hamedan F, Owji L, Rahmanikatigari M, Azadboni TT. Artificial intelligence methods for the diagnosis of breast cancer by image processing: A review. *Breast Cancer: Targets and Therapy*. 2018 Nov 30:219–30, https://doi.org/10.2147/BCTT.S175311.

[2] Jun Y, Craig A, Shafik W, Sharif L. Artificial intelligence application in cybersecurity and cyberdefense. *Wireless Communications and Mobile Computing*. 2021 Oct 27;2021:1–0, https://doi.org/10.1155/2021/3329581.

[3] Avanzo M, Porzio M, Lorenzon L, Milan L, Sghedoni R, Russo G, Massafra R, Fanizzi A, Barucci A, Ardu V, Branchini M. Artificial intelligence applications in medical imaging: A review of the medical physics research in Italy. *Physica Medica*. 2021 Mar 1;83:221–41, https://doi.org/10.1016/j.ejmp.2021.04.010.

[4] Zhao L, Zhu D, Shafik W, Matinkhah SM, Ahmad Z, Sharif L, Craig A. Artificial intelligence analysis in cyber domain: A review. *International Journal of Distributed Sensor Networks*. 2022 Apr;18(4):15501329221084882, https://doi.org/10.1177/15501329221084882.

[5] Barragán-Montero A, Javaid U, Valdés G, Nguyen D, Desbordes P, Macq B, Willems S, Vandewinckele L, Holmström M, Löfman F, Michiels S. Artificial intelligence and machine learning for medical imaging: A technology review. *Physica Medica*. 2021 Mar 1;83:242–56, https://doi.org/10.1016/j.ejmp.2021.04.016.

[6] Meng H, Shafik W, Matinkhah SM, Ahmad Z. A 5g beam selection machine learning algorithm for unmanned aerial vehicle applications. *Wireless Communications and Mobile Computing*. 2020 Aug 1;2020:1–6, https://doi.org/10.1155/2020/1428968.

[7] Houssein EH, Emam MM, Ali AA, Suganthan PN. Deep and machine learning techniques for medical imaging-based breast cancer: A comprehensive review. *Expert Systems with Applications*. 2021 Apr 1;167:114161, https://doi.org/10.1016/j.eswa.2020.114161.

[8] Shafik W, Matinkhah SM, Sanda MN, Shokoor F. Internet of things-based energy efficiency optimization model in fog smart cities. *JOIV: International Journal on Informatics Visualization*. 2021 May 14;5(2):105–12, http://dx.doi.org/10.30630/joiv.5.2.373.

[9] Abdalzaher MS, Elsayed HA, Fouda MM, Salim MM. Employing machine learning and iot for earthquake early warning system in smart cities. *Energies*. 2023 Jan 2;16(1):495, https://doi.org/10.3390/en16010495.

[10] Shafik W, Matinkhah SM, Afolabi SS, Sanda MN. A 3-dimensional fast machine learning algorithm for mobile unmanned aerial vehicle base stations. *International Journal of Advances in Applied Sciences*. 2020;2252(8814):8814: 28–38. https://doi.org/10.11591/ijaas.v10.i1.

[11] Nayak RP, Sethi S, Bhoi SK, Sahoo KS, Nayyar A. Ml-mds: Machine learning based misbehavior detection system for cognitive software-defined multimedia vanets (csdmv) in smart cities. *Multimedia Tools and Applications*. 2023 Jan;82(3):3931–51, https://doi.org/10.1007/s11042-022-13440-8.

[12] Shafik W, Matinkhah M, Sanda MN. Network resource management drives machine learning: a survey and future research direction. *Journal of Communications Technology, Electronics and Computer Science*. 2020 Jul 27;2020:1–5, http://dx.doi.org/10.22385/jctecs.v30i0.312.

[13] Renaud J, Karam R, Salomon M, Couturier R. Deep learning and gradient boosting for urban environmental noise monitoring in smart cities. *Expert Systems with Applications*. 2023 May 15;218:119568, https://doi.org/10.1016/j.eswa.2023.119568.

[14] Shafik W, Matinkhah SM, Ghasemzadeh M. Theoretical understanding of deep learning in uav biomedical engineering technologies analysis. *SN Computer Science*. 2020 Nov;1:1–3, https://doi.org/10.1007/s42979-020-00323-8.

[15] Pandey JK, Veeraiah V, Talukdar SB, Talukdar VB, Rathod VM, Dhabliya D. Smart City Approaches Using Machine Learning and the IoT, In Handbook of Research on Data-Driven Mathematical Modeling in Smart Cities. 2023:345–362. IGI Global, https://doi.org/10.4018/978-1-6684-6408-3.ch018.

[16] Sidiropoulos GK and Papakostas GA, Machine Biometrics – Towards Identifying Machines in a Smart City Environment, 2021 IEEE World AI IoT Congress (AIIoT), Seattle, WA, USA. 2021:0197–0201, https://doi.org/10.1109/AIIoT52608.2021.9454230.

[17] Shafik W. IoT-Based Energy Harvesting and Future Research Trends in Wireless Sensor Networks. Handbook of Research on Network-Enabled IoT Applications for Smart City Services. 2023:282–306, https://doi.org/10.4018/979-8-3693-0744-1.ch016.

[18] Mukhopadhyay S, Suryadevara NK. Smart cities and homes: Current status and future possibilities. *Journal of Sensor and Actuator Networks*. 2023 Mar 16;12(2):25, https://doi.org/10.3390/jsan12020025.

[19] Shafik W. A Comprehensive Cybersecurity Framework for Present and Future Global Information Technology Organizations. In Effective Cybersecurity Operations for Enterprise-Wide Systems. 2023: 56–79. IGI Global, https://doi.org/10.4018/978-1-6684-9018-1.ch002.

[20] Maqsood M, Yasmin S, Gillani S, Bukhari M, Rho S, Yeo SS. An efficient deep learning-assisted person re-identification solution for intelligent video surveillance in smart cities. *Frontiers of Computer Science.* 2023 Aug;17(4):174329, https://doi.org/10.1007/s11704-022-2050-4.

[21] Shafik W. Making cities smarter: IOT and SDN applications, challenges, and future trendsCities Smarter: IoT and SDN Applications, Challenges, and Future Trends. In *Opportunities and Challenges of Industrial IoT in 5G and 6G Networks.* 2023:73–94. IGI Global, https://doi.org/10.4018/978-1-7998-9266-3.ch004.

[22] Chao Z. Machine learning-based intelligent weather modification forecast in smart city potential area. *Computer Science and Information Systems.* 2023, 20(2) 631–656, https://doi.org/10.2298/CSIS220717018C.

[23] Tham JC, Verhulsdonck G. Smart education in smart cities: Layered implications for networked and ubiquitous learning. *IEEE Transactions on Technology and Society.* 2023 Jan 25;4(1):87–95, https://doi.org/10.1109/TTS.2023.3239586.

[24] Shafik W. Cyber Security Perspectives in Public Spaces: Drone Case Study. In Handbook of Research on Cybersecurity Risk in Contemporary Business Systems. 2023: 79–97. IGI Global, https://doi.org/10.4018/978-1-6684-7207-1.ch004.

[25] Javed AR, Ahmed W, Pandya S, Maddikunta PK, Alazab M, Gadekallu TR. A survey of explainable artificial intelligence for smart cities. *Electronics.* 2023 Feb 18;12(4):1020, https://doi.org/10.3390/electronics12041020.

[26] Shafik W. Wearable Medical Electronics in Artificial Intelligence of Medical Things. *Handbook of Security and Privacy of AI-Enabled Healthcare Systems and Internet of Medical Things.* 2024:21–40, https://doi.org/10.1201/9781003370321-2.

[27] Kumar K, Pande BP. Air pollution prediction with machine learning: a case study of Indian cities. *International Journal of Environmental Science and Technology.* 2023 May;20(5):5333–48, https://doi.org/10.1007/s13762-022-04241-5.

[28] Shafik W. A Comprehensive Cybersecurity Framework for Present and Future Global Information Technology Organizations. In Effective Cybersecurity Operations for Enterprise-Wide Systems 2023 (pp. 56–79). IGI Global, https://doi.org/10.4018/978-1-6684-9018-1.ch002.

[29] Gharaibeh AA, Jaradat MA, Kanaan LM. A machine learning framework for assessing urban growth of cities and suitability analysis. *Land.* 2023 Jan 9;12(1):214, https://doi.org/10.3390/land12010214.

[30] Fan C, Xu J, Natarajan BY, Mostafavi A. Interpretable machine learning learns complex interactions of urban features to understand socio-economic inequality. *Computer-Aided Civil and Infrastructure Engineering.* 2023;38(14):2013–2029, https://doi.org/10.1111/mice.12972.

[31] Wang L, Zhao Y, Shi J, Ma J, Liu X, Han D, Gao H, Huang T. Predicting ozone formation in petrochemical industrialized Lanzhou city by interpretable ensemble machine learning. *Environmental Pollution.* 2023 Feb 1;318:120798, https://doi.org/10.1016/j.envpol.2022.120798.

[32] VS DP, Sethuraman SC, Khan MK. Container Security: Precaution levels, Mitigation Strategies, and Research Perspectives. *Computers & Security*. 2023 Sep 22:103490, https://doi.org/10.1016/j.cose.2023.103490.

[33] Javaid M, Haleem A, Singh RP, Suman R. Towards insighting cybersecurity for healthcare domains: A comprehensive review of recent practices and trends. *Cyber Security and Applications*. 2023 Mar 11:100016, https://doi.org/10.1016/j.csa.2023.100016.

[34] Ishrat I, Hasan M, Farooq A, Khan FM. Modelling of consumer challenges and marketing strategies during crisis. *Qualitative Market Research: An International Journal*. 2023 Mar 21, https://doi.org/10.1108/QMR-12-2021-0149.

[35] Padmanaban S, Samavat T, Nasab MA, Nasab MA, Zand M, Nikokar F. Electric vehicles and IoT in smart cities. *Artificial Intelligence-based Smart Power Systems*. 2023 Jan 25:273–90, https://doi.org/10.1002/9781119893998.ch14.

[36] Ajao LA, Apeh ST. Secure edge computing vulnerabilities in smart cities sustainability using petri net and genetic algorithm-based reinforcement learning. *Intelligent Systems with Applications*. 2023 May 1;18:200216, https://doi.org/10.1016/j.iswa.2023.200216.

[37] Mishra A, Jha AV, Appasani B, Ray AK, Gupta DK, Ghazali AN. Emerging technologies and design aspects of next generation cyber physical system with a smart city application perspective. *International Journal of System Assurance Engineering and Management*. 2023 Jul;14(3):699–721, https://doi.org/10.1007/s13198-021-01523-y.

[38] Rawat B, Bist AS, Apriani D, Permadi NI, Nabila EA. Ai based drones for security concerns in smart cities. *APTISI Transactions on Management (ATM)*. 2023;7(2):125–30, https://doi.org/10.33050/atm.v7i2.1834.

[39] Rani P, Sharma R. Intelligent transportation system for internet of vehicles based vehicular networks for smart cities. *Computers and Electrical Engineering*. 2023 Jan 1;105:108543, https://doi.org/10.1016/j.compeleceng.2022.108543.

[40] Mall PK, Narayan V, Pramanik S, Srivastava S, Faiz M, Sriramulu S, Kumar MN. FuzzyNet-Based Modelling Smart Traffic System in Smart Cities Using Deep Learning Models. In Handbook of Research on Data-Driven Mathematical Modeling in Smart Cities 2023. 76–95. IGI Global, https://doi.org/10.4018/978-1-6684-6408-3.ch005.

[41] Aditya A, Anwarul S, Tanwar R, Koneru SK. An IoT assisted Intelligent Parking System (IPS) for Smart Cities. *Procedia Computer Science*. 2023 Jan 1;218:1045–54, https://doi.org/10.1016/j.procs.2023.01.084.

[42] Liu C, Ke L. Cloud assisted Internet of things intelligent transportation system and the traffic control system in the smart city. *Journal of Control and Decision*. 2023 Apr 3;10(2):174–87, https://doi.org/10.1080/23307706.2021.2024460.

[43] Khattak SB, Nasralla MM, Farman H, Choudhury N. Performance evaluation of an IEEE 802.15. 4-based thread network for efficient internet of things communications in smart cities. *Applied Sciences*. 2023 Jun 30;13(13):7745, https://doi.org/10.3390/app13137745.

[44] Caputo F, Magliocca P, Canestrino R, Rescigno E. Rethinking the role of technology for citizens' engagement and sustainable development in smart cities. *Sustainability*. 2023 Jul 1;15(13):10400, https://doi.org/10.3390/su151310400.

[45] El Archi Y, Benbba B, Nizamatdinova Z, Issakov Y, Vargáné GI, Dávid LD. Systematic literature review analyzing smart tourism destinations in context of sustainable development: current applications and future directions. *Sustainability*. 2023 Mar 13;15(6):5086, https://doi.org/10.3390/su15065086.

[46] Zhu, Q., & Zhang, H. (2022). Teaching strategies and psychological effects of entrepreneurship education for college students majoring in social security law based on deep learning and artificial intelligence. *Frontiers in Psychology*, 13, 779669. https://doi.org/10.3389/fpsyg.2022.779669.

[47] Yu, J., Uzuner, U., Long, B., Wang, Z., Yuan, J. S., & Dai, S. Y. (2023). Artificial intelligence-based HDX (AI-HDX) prediction reveals fundamental characteristics to protein dynamics: Mechanisms on SARS-CoV-2 immune escape. *IScience*, 26(4). https://doi.org/10.1016/j.isci.2023.106282.

[48] Yang, Z., Jianjun, L., Faqiri, H., Shafik, W., Talal Abdulrahman, A., Yusuf, M., & Sharawy, A. M. (2021). Green Internet of Things and Big Data Application in Smart Cities Development. *Complexity*, 2021. https://doi.org/10.1155/2021/4922697.

[49] Wu, F., Chen, Y., & Han, D. (2022). Development Countermeasures of College English Education Based on Deep Learning and Artificial Intelligence. *Mobile Information Systems*, 2022. https://doi.org/10.1155/2022/8389800.

[50] Tang, A., Tian, L., Gao, K., Liu, R., Hu, S., Liu, J., Xu, J., Fu, T., Zhang, Z., Wang, W., Zeng, L., Qu, W., Dai, Y., Hou, R., Tang, S., & Wang, X. (2023). Contrast-Enhanced Harmonic Endoscopic Ultrasound (CH-EUS) MASTER: A Novel Deep Learning-Based System in Pancreatic Mass Diagnosis. *Cancer Medicine*. https://doi.org/10.1002/cam4.5578.

[51] Gumbs, A. A., Grasso, V., Bourdel, N., Croner, R., Spolverato, G., Frigerio, I., ... & Elyan, E. (2022). The advances in computer vision that are enabling more autonomous actions in surgery: a systematic review of the literature. *Sensors*, 22(13), 4918.

[52] Jin, X., Wang, X., Cao, X., & Xue, C. (2023). Construction and recognition of acoustic ID of ancient coins based on deep learning of artificial intelligence for audio signals. *Heritage Science*, 11(1), 46..

[53] Shubham, S., Jain, N., Gupta, V., Mohan, S., Ariffin, M. M., & Ahmadian, A. (2023). Identify glomeruli in human kidney tissue images using a deep learning approach. *Soft Computing*, 27(5). https://doi.org/10.1007/s00500-021-06143-z.

[54] Li, W. (2022). Analysis of piano performance characteristics by deep learning and artificial intelligence and its application in piano teaching. *Frontiers in Psychology*, 12. https://doi.org/10.3389/fpsyg.2021.751406.

[55] Jin, X., Wang, X., Cao, X., & Xue, C. (2023). Construction and recognition of acoustic ID of ancient coins based on deep learning of artificial intelligence for audio signals. *Heritage Science*, 11(1), 46. https://doi.org/10.1186/s40494-023-00891-x.

[56] Bastidas, S. E. C., Porras, A. M., Gómez, A. P., García, I. M., & González, L. (2022, November). Elderly, Their Emotions and Deep Learning Techniques to Help Their Dignified and Positive Aging. In *International Conference on Ubiquitous Computing and Ambient Intelligence* (pp. 221–233). Cham: Springer International Publishing.

[57] Ali, R., Balamurali, M., & Varamini, P. (2022). Deep learning-based artificial intelligence to investigate targeted nanoparticles' uptake in TNBC cells. *International Journal of Molecular Sciences*, 23(24). https://doi.org/10.3390/ijms232416070.

CHAPTER 4

Strategies Used in Machine Learning for VANETs

Muhammad Hamza Sajjad, Faisal Rehman, Muhammad Muneer, and Ahmed Ashfaq

4.1 INTRODUCTION

The field of machine learning (ML) in wireless communication networks has had remarkable growth in the past few years due to advancements in science and technology [1]. This improvement had a huge impact on the mobile phone market and Internet Things (IoT) devices. These devices provide different functionalities, including the ability to store and process data, all while connected to the internet. Smartphones are considered next-generation, multi-functional devices that combine the features of advanced smartphones and Vehicular Ad Hoc Networks PDAs [2]. If we look at the data from Google Trends in Figure 4.1, user interest in "mobile phones" exceeded that of "Computers," "Laptops," and "Tablets" in the years 2017 to 2021. People use smartphones for different activities, such as voice communication, e-mailing, instant communication, video conferencing, internet purchasing, entertainment, TikTok, Instagram, Snapchat, etc. [3]. Mobile applications and desktop applications differ in their execution environment due to device constraints and requirements, which can affect functionality and design. Desktop applications are designed for static execution environments because they are usually used in offices or homes and some other static locations, while mobile phone applications need to adjust to dynamic environments and various contexts, which is known as context awareness. Computing shifting towards pervasive and ubiquitous environments and the world around mobile applications is continually changing, requiring them to be adaptable and responsive to their surroundings.

Artificial intelligence (AI) techniques have seen a sharp increase in use in recent years as smart mobile phones have made it possible for gadgets to function intelligently. AI applications for mobile data can use organized, semi-structured, and unstructured data

DOI: 10.1201/9781032648323-4

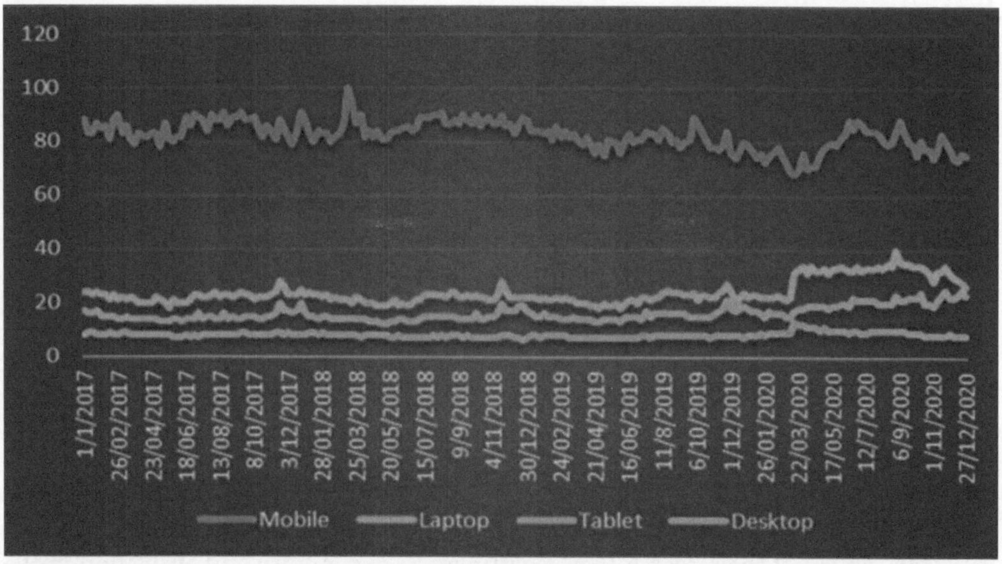

FIGURE 4.1 Trends in user interest over time, shown by the timestamp information on the x-axis and the popularity score on the y-axis, which ranges from 0 to 100 (maximum).

among other types. Artificial neural networks (ANN) [4], expertise systems (ES), knowledge representation, and the use of deep learning and ML approaches are all common AI methodologies. To enable the creation of intelligent apps for mobile devices, the choice of these strategies depends on the data characteristics. AI-based models are frequently employed in numerous intelligent mobile applications in practical applications. The control of the COVID-19 pandemic is one of them. Other examples include customized recommendation systems, AI-powered assistants, mobile company tools, and medical care. A summary of these kinds of apps and how they integrate with AI methods in the context of wireless communication networks is provided in Section 4.6 of the chapter. The broad accessibility of cellphone data and the quick development of data analysis tools have led to a paradigm change in favor of context-aware intelligent computing. Due to the smartphone's capability to recognize the user's present location and context, applications and services related to them are frequently referred to as context-aware. The fact that these applications may dynamically adapt to the user's changing demands qualifies them as adaptive software. Additionally, they are seen as intelligent since they make use of based on data AI models, which allow them to intelligently assist clients in various day-to-day circumstances based on their needs.

The importance of artificial intelligence (AI) in the creation of mobile apps, particularly in areas of mobile phone data analysis and smart application within the context of wireless communication networks, is the main topic of this chapter. In diverse application situations, we want to investigate how AI approaches might be applied to develop data-driven, smart mobile apps that can improve people's lives. Our main goal is to operate as a fundamental resource for academic and professional researchers and practitioners who seek to explore and develop intelligent smartphone applications powered by AI, taking into account the special features of these apps compared to conventional ones within the domain of wireless communication networks.

This chapter includes:

- A synopsis and description of the mobile data analysis paradigm, a necessary step in creating data-driven, intelligent apps. To achieve this, we quickly review pertinent techniques and frameworks to stimulate more in-depth study in this area within the context of wireless communication networks.
- Representations of knowledge and rule-based expert systems, along with a range of MI and DL techniques to create AI-based modeling for intelligent mobile apps in wireless communication networks.
- An examination of the efficacy of various AI-powered smart apps across a range of application sectors, highlighting the significance of AI-based modeling in enhancing human lives.
- A discussion of the research that is pertinent to our inquiry and evaluation in the field of smartphone data analysis or smart applications.

This chapter is organized as follows:

- The scope of our research and its introduction is covered in Section 4.2.
- The history of conventional data science, context-aware computing for mobile devices, and related research in the area of information-driven mobile systems and services are covered in Section 4.3.
- The idea of mobile data science methodology is covered in Section 4.4.
- We outline an AI-based modeling strategy that adheres to the parameters of our investigation in Section 4.5.
- The many intelligent apps that make use of AI technology within the realm of wireless communication networks are compiled in Section 4.6.
- Some research issues and possible future paths are covered in Section 4.7.
- We highlight several important topics relating to our study in Section 4.8.
- Finally, Section 4.9 concludes the chapter.

4.2 PURPOSE AND RANGE OF THE CHAPTER

Information is transmitted over space using electromagnetic waves in wireless communication network technology. It includes encoding data on radio waves, which are modulated and delivered via antennas before demodulating at the other end. This technology depends on essential elements including Wireless Network Virtualization (WNV), Information-Centric Networking (ICN), Device-to-Device (D2D) connectivity, and Software-Defined Network (SDN) [23]. The communication-cache system, which improves network performance by storing frequently used data locally, is a typical feature. Figure 4.2 shows an example system architecture.

This chapter also serves as inspiration for our chapter, which explores the application of AI and mobile data analysis in intelligent models and smartphone apps. In the connected world of today, where phones and Internet of Things (IoT) gadgets have proliferated and are widely acknowledged to have a key role, our chapter is especially pertinent. To create intelligent algorithms and apps that operate within wireless communication networks, we, therefore, intend to examine how mobile data analysis and ML can effectively collaborate. In this section, we also describe the size of our investigation.

Artificial intelligence (AI), cloud computing, data science (DS), machine learning, and cybersecurity are the technologies used in the context of today's fast-evolving technological landscape [5]. The trend towards pervasive and context-aware intelligent computing is rising as more people utilize computing devices like mobile phones and related apps outside of the desktop environment. Effectively processing mobile data and enhancing app behavior by incorporating pertinent contextual data, such as temporal, geographic, social, environmental, and device-related contexts, continues to be a significant challenge in this evolving research field. Typically, when we talk about context, we mean any details that describe a scenario.

In the context of ML in wireless communication network mobile applications can be made intelligent by using a variety of techniques such as artificial neural networks or ML,

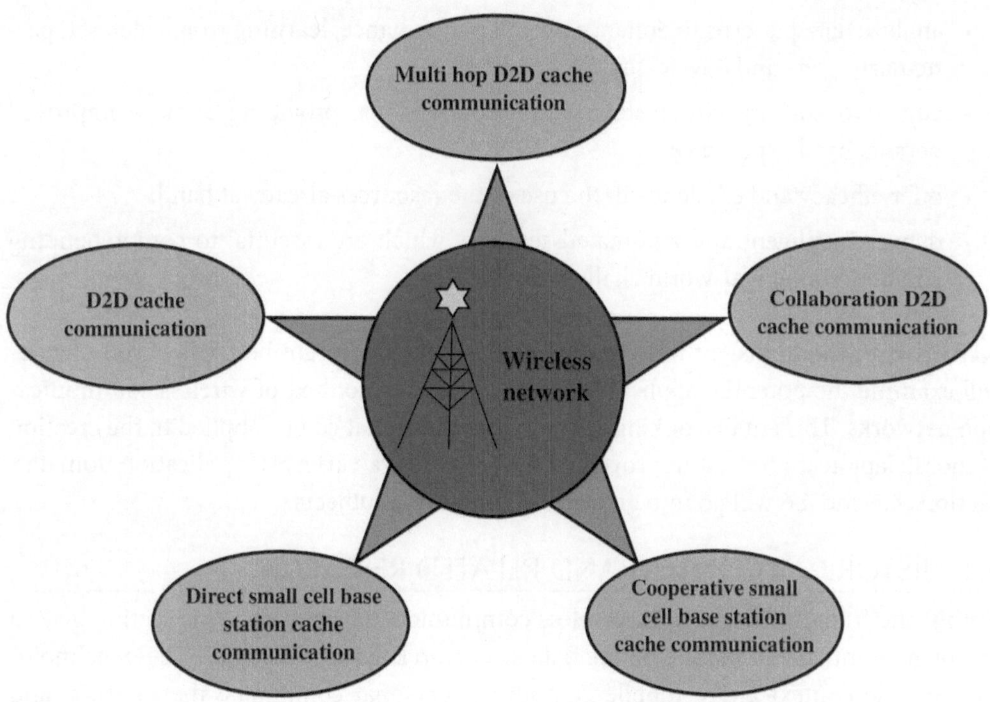

FIGURE 4.2 A typical wireless communication network's schematic design of a communication cache system.

natural language processing (NLP), rule-based expert systems, and knowledge representation, suited to the precise data requirements and features. For example, algorithms using ML often uncover insights or inherent patterns in the data from phones to improve forecasts and judgments in smart systems [6]. We can tackle hard problems even while employing a variety of data sets thanks to deep learning [7], a branch of ML. NLP plays a vital role in AI by enabling the extraction of insights from unprocessed mobile content expressed in languages like English or Hindi. This process involves analyzing and understanding human languages, such as English or Hindi, to derive valuable intelligence. A representation of information and a rule-based expert system are two additional crucial components of AI that we examine in our research. Expert systems (ES) are intelligent systems created to handle complicated issues by arguing based on knowledge, mainly via IF-THEN rules rather than ordinary conventional programming instructions. ES often mimics the decision-making abilities of a human expert within the wireless communication network context.

The type of relevant data and the AI tasks used to create an effective model are therefore crucial to the success of AI-based mobile applications. This chapter will examine the numerous justifications for why AI applications are crucial in mobile systems and applications. In summary, AI can:

- enable the development of the smartphone industry by turning mobile applications into smart software that can foresee the future and make decisions based on user requirements.
- analyze data patterns to enhance overall performance, learning from a dataset, particularly user- and device-centric situations.
- adjust to shifting circumstances and innovations, providing a more improved personalized experience.
- offer efficacy and efficiency in the use of the resources already at hand.
- deliver intelligent and automated services, which are essential to comprehending and addressing real-world challenges.

To give a general insight into how AI-based modeling might be applied, this chapter will examine the potential application of AI within the context of wireless communication networks. This entails looking at several methods that can be applied to the creation of mobile applications that improve human life across a variety of application domains. Sections 4.5 and 4.6 will go into greater detail on these subjects.

4.3 HISTORICAL CONTEXT AND RELATED RESEARCH

Within the framework of ML in wireless communication networks, this section gives a thorough summary of mobile phone data science-related technologies. These technologies include context-aware mobile computing, personal computers, the Internet, and classical data science, all of which are pertinent to our chapter.

4.3.1 Heterogeneous Networks in Wireless Communication

A fundamental design for the creation of wireless communication networks is heterogeneous networks. To improve the reach of macro cell stations, mini cell stations are often used constrained spaces. A radio access network (RAN) is still needed to route data to small cellular base station buffers. While device-to-device (D2D) communication technology enables users to exchange materials, it is more expensive, has a smaller storage capacity, and has a smaller battery life than tiny cellular base stations. As a result, the framework primarily addresses the common paradigms of D2D, multi-hop D2D, cooperative D2D, direct small cellular base stations, and small cellular base stations. The primary idea behind the D2D cache-communication paradigm is that user equipment may improve the quality of user service across different cells by getting material from nearby user equipment through D2D communication. The multi-hop D2D cache communication approach primarily relies on using neighboring user devices as a relays to obtain material from the tiny cellular base stations in other cells. This method enables more widespread cooperative caching and distribution while also enhancing resource usage. A copy of the material is cached on many user devices, and this is referred to as the collaborative D2D cache-communication paradigm. When the necessary content is not pre-cached in the nearby device, the user can also directly request the content from the linked small cellular base station, resulting in shorter latency. This is known as the direct small cellular base station cache-communication paradigm. According to the cooperative cache-communication model for small cellular base stations, a virtual link is created with the user's device to acquire required content from neighboring small cellular base stations if the desired content is not cached in a nearby small cellular base station.

4.3.2 Data Science

We are in the midst of a data revolution [8], where making intelligent decisions based on data is crucial. Various technologies such as ML, data science, neural networks, and advanced analytics play a significant role in this data-centric approach. Researchers commonly refer to the interdisciplinary fields of data collection, preprocessing, inference, and decision-making as data science. It encompasses the application of diverse scientific methods, processes, and systems to analyze and understand real-world phenomena. Data science integrates various disciplines such as statistics, informatics, information technology, management, and social science to examine and understand its environments. By employing a data-to-knowledge-to-wisdom approach and method, data science aims to convert data into valuable understanding and informed decisions [8]. In essence, it can be defined as the scientific exploration of data, resulting in solutions driven by data for the challenges we face.

4.3.3 Network Connectivity and Computers

The growth of the contemporary digital age has been significantly influenced by the development of smartphone computing and the Internet. The most common computing device used by individuals in the real world is a mobile smartphone because of the Internet.

People now connect around the globe for a variety of reasons primarily using mobile devices. While there are many different types of mobile phones in reality, in this chapter, we only refer to cell phones or other smartphone devices with computation and Internet access capabilities. To improve access to data through intelligent computing and the right usage of the tools for the advantage of users, these devices include several important features. Additionally, the field of transportation systems has changed as a result of the convergence of AI and Machine Learning (ML) within Vehicular Ad Hoc Networks (VANET). Vehicle-to-vehicle communication is made possible by the integration of AL/ML algorithms into VANET, which improves traffic control, road safety, and overall vehicle efficiency.

4.3.4 Context-Aware Mobile Computing

In many diverse fields, including ambient intelligence, pervasive computing, and human-computer interaction, the concept of the environment has been extensively used. Initial studies into context-aware computing focused on defining context as the position of people and substances in the setting of mobile and ubiquitous computing [9]. Different types of contexts are considered to achieve specific objectives, such as locational context, user actions [9], temporal data, contextual data, user identification, or social context [10]. Moreover, context takes into account the surrounding information related to the application's circumstances. Schilit et al. [11] identified three key factors in context: (i) location, (ii) social companionship, and (iii) a shared understanding of context that extends beyond mere resource accessibility It defines context as including any data that may be used to characterize an entity's external environment. An entity in this sense might be a person, a place, or an item that has an impact on how a user interacts with an application. This definition considers the program itself in addition to the user. Context can also be seen as a specific type of information that can be utilized to adapt an application.

Context awareness may be the guiding principle of ubiquitous computing, based on the contextual data previously stated. Context-aware computing refers to the capacity to perceive the immediate physical surroundings and adjust application behavior. Context-aware applications may generally adapt to the movement of cellphone users. As a result, context awareness just serves to illustrate how dynamic programs are. Thus, in a ubiquitous computing environment, the amount of user input and the focus required for an app to cater to the user's needs can be reduced through the usage of contextual data within mobile applications. In an earlier section, we briefly reviewed some applications that may be affected differently by various settings.

4.3.5 Mobile Devices and Services

Mobile data gathered from multiple sources and used in research is frequently adapted for use in certain apps, which can differ from one instance to another. Numerous studies on mobile devices, services, and sources of data have been conducted. For instance, call records [12] that record a user's phone calls and provide context information. Smart mobile phones also capture various sorts of contextual information and call-related data, including the user location and social connections between the callers specified by the person's specific telephone number. Smartphone short message services include all messages, such as wanted and unwanted messages sent via text or desired and undesired content, along with their associated data, including user ID, timestamps, and additional information, which may be helpful in the process of automatically separating text message spam [12] or predicting the best or worst times to deliver certain messages. App usage logs include a variety of contextual data, including the date, timestamps, battery status, profile mode (including general, silent, meeting, outdoor, offline), charging status (e.g., complete, charging, or disconnected), geographic location (e.g., workplace, home, on the move, etc.), and other app-related information. Contextual details included in mobile phone data in the notification log include notification type, user location (whether they are at home, work, or elsewhere), user activity (such as whether they are stationary, walking, running, biking, or in a vehicle), date, timestamps, the user's interaction with these alerts (whether they were dismissed or accepted), as well as additional metadata related to the notifications. Weblogs encompass a range of activities such as web searching, email, entertainment, chat, miscellaneous news, TV, internet usage, travel, sports, and banking. They also include relevant contextual information such as date, time, and whether it's a weekend or weekday. Similarly, every mobile phone user's gaming record contains information about the games they've played, along with contextual data like date, time, and day of the week [36]. Furthermore, incorporating ethical considerations such as those found in ethical AI is essential to guaranteeing responsible and safe AI-driven features in smartphones. At the same time, adding power electronics expands the devices' functionality, guaranteeing peak performance and improved user experiences.

Mobile phones also include data related to user telephone calls, SMS app use (e.g., Imo, Facebook Messenger, Skype, etc.), physical movements, Bluetooth and Wi-Fi signals in user's proximity, geolocation data, etc. Numerous apps, including intelligent mobile notification management, intelligent mobile communication, and context-sensitive mobile recommendation, are used. Common applications in the field of mobile data analysis and application include intelligent mobile conversation, efficient smartphone notification control, intelligent mobile suggestions, and more. One of the most appealing and extensively researched applications is intelligent context-aware mobile communication (e.g., smartphone call interruption management). Numerous studies (e.g., [13]) have been conducted on managing smartphone notifications. Similar to this, other studies (e.g., [14]) have been conducted on recommendation systems.

In the field of mobile analytics, a number of methodologies are utilized in a variety of applications, including disruption management, activity recognition, recommendation systems, mobile commerce, etc. For instance, Seo et al. [15]. PYP configuration manager for cell phones is context-aware makes a case for an intelligent interruption management system that employs decision trees. A customized hybrid recommender system was created using the classification technique of Bozanta et al. [16], [17] and Lee et al. [18]. In their interruptibility prediction and management system, Turner et al. [19] and Fogarty et al. [20] employ the categorization technique. In his context-aware mobile applications,

Bedogni et al. [21] apply categorization approaches in the field of transportation. Tan et al. [22] study the use of the multilayer perceptron model for mobile learning. Paireekreng et al. [23] built a recommendation system for personalized mobile games. Additionally, in the field of supervised learning, regression methods like linear regression, support vector regression [24], and collective learning methods like random forest learning are widely used.

In addition to the methods outlined above, some researchers [14] employ association rules to create diverse context-aware mobile services tailored to the demands of consumers. Several studies have used the clustering approach for various reasons in their research. Additionally, significant research [22] is being done on deep learning techniques for a variety of uses in the field of mobile analytics. Another crucial subject to focus on in this field is context engineering, which includes techniques like principal component analysis or context correlation analysis [25]. Table 4.1 provides a summary of the most widely used methods and data-driven tasks.

The community's changing expectations in reaction to current events are what motivate researchers to build intelligent apps through thorough AI-based modeling, which uses some of kinds of cell phone data and methods previously discussed. In this chapter, we concentrate on mobile DS and related smart apps within the context of ML in wireless communication networks. AI techniques, such as deep learning and ML, NLP, knowledge representation, and expert systems can be used to create intelligent mobile apps within wireless communication networks.

4.4 PARADIGM OF MACHINE LEARNING IN VEHICULAR AD HOC NETWORKS

In this section we provide an overview of smartphone data analysis of ML in Vehicular Ad Hoc Networks (VANET).

4.4.1 Knowledge of Mobile Data

Data availability is crucial in the field of smartphone and information-driven smart apps. Mobile datasets often include a variety of contextual elements and details. Understanding the properties of smartphone data, which include a variety of feature types and contextual data, is vital. Unprocessed information from appropriate sources may be used to evaluate different trends and gain insightful knowledge. This information may be used to build a model based on data to help us achieve our goal. Datasets including call records from phones, app logs, websites, etc., are a compilation of user-diverse activities and past context-relevant data [26]. Data may also be gathered from smart city enterprises, health care apps, privacy tools, and sensors used on smartphones and apps. Depending on the application, smart apps depend on the insights gleaned from relevant datasets. We list a few characteristics of smart apps in Table 4.1.

TABLE 4.1 An Overview of the Data-Driven Activities and Methods Applied to Various Context-Aware Mobile Services and Systems

Activities and Methods	Purposes and Objectives	References
Clustering	Segmentation based on time, Modeling of time-series data.	[25]
Association	Management of interruptions, Modeling of usage patterns, Management of notifications, Analysis of recent events.	[13]
Context Engineering	Usage of applications, Analysis using principal component analysis, Analysis of correlation in context, Pre-modeling of context.	[27]
K-Nearest Neighbor	Search on mobile devices, System for recommending content, Prediction of location, Recognition of activities.	[28]
Logistic Regression	Recognition of activities, Modeling of user behavior, System for personalized recommendations, Analysis of health data, Prediction of interruptibility.	[28]
Naive Bayes	Prediction of phone calls, Prediction of location, Management of interruptions.	[29]
Decision Tree	System aware of context, Mobile service provision, Management of interruptions, Prediction of phone calls.	[28]
AdaBoost	Management of interruptions, Prediction of incorruptibility, Prediction of location, System for recommending content.	[29]
Neural Network	Modeling of smartphone power consumption, Mobile payment using credit cards, Mobile learning applications, Characterization of smartphones.	[30]
Random Forest	Prediction of call availability, Instant messaging services, Transportation system management, Recognition of activities.	[29]

4.4.2 Features of Smart Applications

AI and data analytics are the fundamental building blocks of smart apps, which provide personalized and flexible customer experiences. In general, smart apps have the following characteristics:

- *Action-Oriented*: Smart apps analyze user behavior and apply predictive analytics to give individualized and useful results.

- *Responsive in Nature*: Smarts apps must be flexible in order to quickly update their knowledge of their surroundings to create a rewarding user experience.
- *Decisive and Suggestion-Focused:* Smart app have the ability to make judgments and generate recommendations based on user requirements and tastes.
- *Operation across Platforms*: Smart apps must have the ability to provide a consistent user experience across different platforms.

Next, we discuss smartphone data science and AI, which allow smart apps to leverage these qualities.

4.4.3 Artificial Intelligence and Mobile Data Science

Smart apps are all about smartphone data. According to a recent study based on deep learning techniques, the majority of smartphone apps still use manually defined algorithms or custom-written rules like signatures for their various applications [31]. The fundamental disadvantage of these specifically created rule-based techniques is that users must manually construct and maintain the information or rules the apps use. While methods based on rules have benefits in some circumstances, they require a lot of manual effort to stay aligned with the evolving user context environment. But data science, which uses AI methods like deep learning, ML, language processing, representation of knowledge, and trained systems to acquire knowledge and make wise decisions, can be used for collecting information, data analysis, data cleaning, data preprocessing, data visualization, and determination. AI is used to make intelligent computer programs.

Within the framework of wireless communication networks, the key outputs of mobile data products, such as an AI model powered by data, future mobile services, and recommendations, or related smart mobile apps, are the main outcomes of mobile data science. We discuss AI-powered smart apps in in Section 4.6.

4.4.4 Ethical Perspectives on Mobile Security and Privacy

Mobile safety and privacy may be another aspect of smartphone data science data-driven security solutions, even considering smart applications from the angle of AI inside the limits of our study stated above. The majority of people have smartphones for both personal and professional use. Therefore, mobile security and privacy are essential for smart apps. Phones gather and process sensitive data, to which accessibility must be restricted to safeguard the user's integrity or a company's proprietary information. In our study, Sarker et al. [5], discussed security analytics and an AI-based cyber security modeling employing ML approaches. Moreover, AI or ML interaction in VANETs can be used for strengthening mobile security protocols and guaranteeing strong safety standards. Because of this, security insight and data-driven intelligent solutions may be able to effectively identify and counteract these types of mobile security threats.

4.4.5 Ethical Implications of AI in VANETs

Defense and assault are the two facets of wireless security. Attacks relate to procedures where an attack, such as jamming or eavesdropping, is proactively executed to secure future communications. Defensive measures include encryption and similar technologies. Such security-related research provides not only the examination of system weaknesses but also the destabilization of an adversary system's capabilities.

The rapid speed of ML research might eventually lead to every device having some level of intelligence that could be exploited either positively or negatively. The security of the many devices that coexist in the same environment is at risk if such capabilities are present in malevolent devices (i.e., those that intend to purposefully create interference).

Typically, game theory or optimization frameworks have been used to solve these issues. They provide useful insights, but often they presuppose static surroundings, static action spaces for an enemy, etc., which may not be the case if the adversary has ML capabilities. As a result, we must research these systems from both an attack and defense [32] standpoint. We must create ML models that can quickly adapt to the surrounding environment and prevent the adversary from communicating or interfering. From a defensive standpoint, we must create a communication system that is resistant to all attacks, and we can do this by using ML methods.

Furthermore, strengthening a communication network's resistance to interference and possible attack requires the incorporation of power electronics into wireless security procedures. Simultaneously, implementing ethical AI principles guarantees the conscientious application of intelligent systems in wireless contexts, respecting ethical norms while utilizing AI technology for security.

4.5 MODELING & ETHICAL CONSIDERATIONS IN AI FOR MOBILE SERVICES AND VANETS

As discussed, smartphone data science uses a variety of AI techniques to determine smart choices in smartphone apps or operations. As part of our analysis, we categorize AI techniques into some groups, including fundamental ML and algorithmic techniques for deep learning, language processing, representation of knowledge, and trained systems. These AI-based techniques are discussed as follows.

4.5.1 Mobile Data and Machine Learning Modeling

Artificial intelligence including ML as well as deep neural network learning can enable mobile devices to make decisions without human intervention. For example, AI algorithms can be used to examine user behavior patterns in phone log data to create tailored recommendations and suggestions [33]. Supervised learning is used to create a model from gathered data when goals are specified to be reached from a particular collection of features [6]. This involves using algorithms such as adaptive boosting, logistic regression, Naive Bayes, decision trees, support vector machines, K-nearest neighbors, etc., to categorize or predict the outcome of a future event. These methods of classification can be used to create prediction models for everything from the upcoming usage of phones to mobile malware attack prediction. The effectiveness of the resulting predictive model can be increased by doing various feature design tasks, including selecting features, extraction, etc., information from analysis [27]. Unsupervised learning, on the other hand, examines similarities across unlabeled data without labeling or classifying the data. Depending on the data characteristics, a variety of clustering techniques, including K-means, hierarchical clustering, mean shift, density peak clustering, and spectral clustering, can be used. A rule-based ML model can be built for smartphone users using association rule learning approaches like AIS, FP-Tree, RARM, and ABC-RuleMiner. Many deep neural learning techniques, including convolutional neural networks, recurrent neural networks, long-short term memories [34], multilayer perceptrons, etc., which were derived from ANN, can also be used in the learning process. These deep learning models allow for the inclusion of several hidden layers to finish the process. These AI-driven modeling approaches enable mobile devices to analyze data and offer intelligent solutions.

Due to their capacity to learn from data, models based on ML have the potential to revolutionize the market for mobile applications. Deep neural networks and other methods of ML can be widely used to make mobile platforms more user-friendly, improving the user experience and facilitating the creation of smart apps.

4.5.2 Extraction of Natural Language for Smartphone Applications

An essential area of ML, natural language processing (NLP), focuses on how machines and human beings interact naturally. One of NLP's primary objectives is to create intelligence from unprocessed data or information familiar to the speaker, like British English or Hindi. Utilizing every language's grammar, syntax, and traditions, NLP approaches allow machines to process and comprehend spoken and written words in a variety of languages. These methods make it possible to do tasks such as analyzing text, transcribing voice, deciphering significance, gauging emotions, collecting opinions, and eventually recognizing key elements in a smart system [35]. NLP can also be used to extract emotions linked to either favorable or unfavorable polarities for particular subjects identified in a piece of writing. This approach analyzes the text using NLP methods to determine what emotions are related to each issue. Thus, when there is unprocessed mobile information accessible, Ml can play a vital role in creating smart applications and be a key component of our research.

Text-based content is now widely read on cell phones. NLP approaches use text analysis to determine patterns and organize information from text content to make applications more knowledgeable and smarter. For example, NLP can be used for document summarization to avoid having to scroll through pages and pages of text.

4.5.3 Ethical Frameworks in VANETs: Privacy and Security Implications

AI is used by VANETs to improve safety and communication in transportation networks. However, there are moral dilemmas with using AI in VANETs, especially with regard to security and privacy. Thus, it is imperative to put ethical frameworks in place to handle these security and privacy issues and optimize the advantages of AI-driven technologies. Pseudonymization strategies, encryption mechanisms, and secure communication routes are all necessary to preserve user data privacy on VANETs. It is also critical to keep strong security mechanisms in place within VANETs in order to thwart malicious attempts and guarantee secure vehicle communication. AI-powered security solutions are developed with ethical frameworks in mind, with the goal of minimizing risks and defending against attacks while respecting moral standards and user confidence.

4.6 PERSONALIZATION

Today, the majority of smartphone applications heavily depend on personalization based on a user's past data. By considering behavior, relationships, or tastes, AI models may efficiently uncover insightful information from customer data, which can then be applied to provide personalization within apps. For example, an actual-life app built on the principles found can be a smart telephone interrupt application that manages calls that arrive instantly by a specific user's actions [37]. Additionally, real-world examples of personalized services offered to smartphone users include smartphone alert management, app usage forecast and administration, etc.

4.6.1 Recommendation for Mobile

A recommendation system can be created with ML to forecast a user's desires and inclinations by examining their behaviors and the actions of other users, therefore producing tailored suggestions, as described in Ref. [38]. Smartphone recommendation systems are based on data from users and can provide relevant information the consumer is searching using ML algorithms. These systems minimize user search time by providing relevant options based on user data. Examples include guidebooks for tourists [39], food or restaurant recommendations, flight and travel suggestions, product recommendations based on user reviews, etc.

4.6.2 Smartphone Business

Smartphone applications could improve efficiency and effectiveness in both the business-to-customer and business-to-business fields. Enterprise is undergoing an exciting shift as a result of the growing accessibility and development of AI and ML, which is seen as the new electronic frontier for organizations. Most businesses prioritize client service, prompting contemporary entrepreneurs to focus on establishing smart connections within smartphone apps to cater to consumer needs.. Companies can use information gathered from a variety of sources, including PoS systems, internet usage, cell phones, etc., to determine and effectively enhance user experience. AI methods identify patterns in data and modify the applications themselves to produce more interesting and contextually greater opportunities for user engagement. For example, AI systems can offer customers product suggestions based on past purchases, identify credit card fraud, and provide search results. Moreover, an information-driven smartphone expert system can make smart choices by taking into account a set of business IF-THEN criteria. In general, business-related AI smartphone apps aid in corporate growth, the introduction of new goods and services, the identification of customer needs, and the maintenance of a dominant position in the worldwide marketplace.

4.6.3 Revolutionizing Healthcare Delivery

By streamlining their bodily interactions, smartphone medical apps are improving possibilities for patients, doctors, and connected organizations. These mobile applications can offer access to a range of healthcare services, medical evaluations, medication advice, e-prescriptions, recommendations for basic safety measures, remote health monitoring, and efficient patient care administration in hospitals. Various types of medical health records can be gathered from diverse origins on a vast range of health subjects to examine evaluate and bolster healthcare facilities or develop healthcare information systems (HMISs) [40]. Intelligent health services can be offered with the aid of AI techniques, including ML-based models, which have the potential to save lives by providing personalized medications and preventative measures through data collection. Additionally, AI-powered smartphone applications may be used to find medical providers, schedule appointments, provide medication reminders, etc.

In general, AI-modeling-based smartphone medical care could open up new opportunities for all users, but especially those residents in nations with low incomes who can receive high-quality medical services via wireless communication networks.

4.7 RESEARCH QUESTIONS AND SUGGESTED NEXT STEPS

The rapid advancement of mobile gadgets, the IoT, and AI techniques poses a fundamental challenge: analyzing vast amounts of relevant data gathered from multiple origins and extracting actionable information. Thus, in this part, we describe the following challenges:

- The main requirement for working in the field of smartphone data science is a reference dataset. As a result, gathering real-world information relevant to a given application, such as classified, numerical, or textual data is the first step in creating smart mobile applications. For example, to control cellphone interruptions, it is necessary to analyze both an individual's behavioral data and pertinent contextual information [41]. Patient data and the associated contextual data may also be helpful for smart medical care. Therefore, it is crucial to integrate and effectively manage mobile data to make it easier to extract trustworthy knowledge from the data by using AI methods.
- Another challenge is accurately simulating smartphone users and their behaviors. The primary objective of smartphone user modeling is to customize and adapt solutions to the unique requirements of users. The system must deliver the correct result in the correct context at the correct time [41]. Consequently, several factors, including contextual dependence, user conduct and choices in various settings, must be taken into consideration to create successful user modeling and smart

applications. Building a smart app requires taking into account a variety of settings, including time-based, geographical, social, etc. [42]. The main difficulties are preparing the data, and the identification of practical insights [41].

- Real-world smartphone datasets may have a lot of characteristics or complicated settings, which can lead to several problems like increased complexity of models, over-fitting concerns, and decreased performance of the resulting AI-based model. The choice of characteristics or representation of data significantly impacts how well AI, especially ML algorithms, performs. When unimportant characteristics or contextual information are present in the data, the model learns based on those extraneous features, which reduces the model's performance.
- Another difficulty is obtaining reliable and relevant details from semi-structured or unorganized data on smartphones and tablets. Much written material is routinely consumed on these kinds of devices, including electronic mail, websites, and papers. Therefore, rather than the laptop, and computers, the issue of data overflow is caused by the tiny display of the gadgets. Successfully mining texts or data while taking these factors into account could thus be a further topic in the field of smartphone data science and smart apps.
- To provide suggestions or make judgments, smartphone expert systems require expert knowledge, which is mainly expressed as IF-THEN rules. Large-scale rule-based system development, however, may confront several difficulties. For example, building such systems might be challenging due to the complexity of reasoning itself. Lightweight rule-based inference engines are currently lacking, making it difficult to reason on mobile devices. So in the case of an expert system based on rules for mobile devices, it would be advantageous in terms of result and convenience to have a set of simple rules. Ontologies [43] offer a versatile and comprehensive means of representing high-level concepts and relationships among attributes, allowing for the modeling of intricate dependencies within a specific problem domain. By utilizing ontologies, both the system and the user can utilize a shared taxonomy, which plays a vital role in developing an expert system. Ontological modeling and reasoning can be helpful in this situation. As a result, another study topic could be the development of an appropriate ontology or knowledge representation model for addressing.
- Extremely dynamic computing environments result from the portability of users, apps, and computing devices like smartphones. Apps for ubiquitous computing can adapt to fluctuations in available resources, such as user situations, network access, etc., in contrast to desktop apps, which depend on a well-prepared, mainly static set of resources. To complete activities for consumers, they are usually required to collaborate spontaneously with previously unidentified software services. Software for widespread computing must therefore be flexible and adaptable. As an illustration, an app may need to change its output style to be less invasive when switching from an office setting to a moving car [41].

4.8 CONCLUSION

AI modeling combines knowledge representation, expert systems, DL, ML, and NLP. This chapter briefly discussed how such AI modeling might be utilized to develop smart mobile apps, including personalized suggestions for medical services. An effective mobile smart system needs the appropriate AI-based modeling to align with the data's properties. The complex algorithms must then be trained using information and data gathered specifically for the targeted application before the system can offer ideas and assistance with choices to the users. As discussed, there are research challenges and next steps within the field of smartphones and smart applications to improve the landscape of the use of ML in wireless communication networks.

REFERENCES

[1] M. Peng, G. Zeng, Z. Sun, J. Huang, H. Wang, and G. Tian, "Personalized app recommendation based on app permissions," *World Wide Web*, vol. 21, pp. 89–104, 2018.

[2] P. Zheng and L. M. Ni, "Spotlight: the rise of the smart phone," *IEEE Distributed Systems Online*, vol. 7, no. 3, p. 3, 2006.

[3] M. A. El Khaddar and M. Boulmalf, "Smartphone: the ultimate IoT and IoE device," *Smartphones from an Applied Research Perspective*, vol. 137, 2017.

[4] H. Sharif *et al.*, "Application of artificial neural networks in satellite imaging–A systematic review," in *2023 4th International Conference on Computing, Mathematics and Engineering Technologies (iCoMET)*, IEEE, 2023, pp. 1–5.

[5] I. H. Sarker, A. S. M. Kayes, S. Badsha, H. Alqahtani, P. Watters, and A. Ng, "Cybersecurity data science: an overview from machine learning perspective," *J Big Data*, vol. 7, no. 1, pp. 1–29, 2020.

[6] I. H. Sarker, A. S. M. Kayes, and P. Watters, "Effectiveness analysis of machine learning classification models for predicting personalized context-aware smartphone usage," *J Big Data*, vol. 6, no. 1, pp. 1–28, 2019.

[7] H. Sharif, F. Rehman, A. Rida, and A. Sharif, "Segmentation of images using deep learning: A survey," in *2022 2nd International Conference on Digital Futures and Transformative Technologies (ICoDT2)*, IEEE, 2022, pp. 1–6.

[8] L. Cao, "Data science: a comprehensive overview," *ACM Computing Surveys (CSUR)*, vol. 50, no. 3, pp. 1–42, 2017.

[9] B. N. Schilit and M. M. Theimer, "Disseminating active map information to mobile hosts," *IEEE Netw*, vol. 8, no. 5, pp. 22–32, 1994.

[10] D. Franklin and J. Flaschbart, "All gadget and no representation makes jack a dull environment," in *Proceedings of the AAAI 1998 spring symposium on intelligent environments*, 1998, pp. 155–160.

[11] B. Schilit, N. Adams, and R. Want, "Context-aware computing applications," in *1994 First Workshop on Mobile Computing Systems and Applications*, IEEE, 1994, pp. 85–90.

[12] N. Eagle and A. Pentland, "Reality mining: sensing complex social systems," *Pers Ubiquitous Comput*, vol. 10, pp. 255–268, 2006.

[13] A. Mehrotra, R. Hendley, and M. Musolesi, "PrefMiner: Mining user's preferences for intelligent mobile notification management," in *Proceedings of the 2016 ACM International Joint Conference on Pervasive and Ubiquitous Computing*, 2016, pp. 1223–1234.

[14] H. Zhu, E. Chen, H. Xiong, K. Yu, H. Cao, and J. Tian, "Mining mobile user preferences for personalized context-aware recommendation," *ACM Transactions on Intelligent Systems and Technology (TIST)*, vol. 5, no. 4, pp. 1–27, 2014.

[15] I. H. Sarker, A. S. M. Kayes, M. H. Furhad, M. M. Islam, and M. S. Islam, "E-miim: an ensemble-learning-based context-aware mobile telephony model for intelligent interruption management," *AI Soc*, vol. 35, pp. 459–467, 2020.

[16] A. Bozanta and B. Kutlu, "Developing a contextually personalized hybrid recommender system," *Mobile Information Systems*, vol. 2018, 2018.

[17] S. Ashraf, F. Rehman, H. Sharif, H. Kim, H. Arshad, and H. Manzoor, "Fake reviews classification using deep learning," in *2023 International Multi-disciplinary Conference in Emerging Research Trends (IMCERT)*, IEEE, 2023, pp. 1–8.

[18] W.-P. Lee, "Deploying personalized mobile services in an agent-based environment," *Expert Syst Appl*, vol. 32, no. 4, pp. 1194–1207, 2007.

[19] L. D. Turner, S. M. Allen, and R. M. Whitaker, "Interruptibility prediction for ubiquitous systems: conventions and new directions from a growing field," in *Proceedings of the 2015 ACM international joint conference on pervasive and ubiquitous computing*, 2015, pp. 801–812.

[20] J. Fogarty, S. E. Hudson, C. G. Atkeson, D. Avrahami, J. Forlizzi, S. Kiesler, J. C. Lee, J. Yang, "Predicting human interruptibility with sensors," *ACM Transactions on Computer-Human Interaction (TOCHI)*, vol. 12, no. 1, pp. 119–146, 2005.

[21] L. Bedogni, M. Di Felice, and L. Bononi, "Context-aware Android applications through transportation mode detection techniques," *Wirel Commun Mob Comput*, vol. 16, no. 16, pp. 2523–2541, 2016.

[22] G. W.-H. Tan, K.-B. Ooi, L.-Y. Leong, and B. Lin, "Predicting the drivers of behavioral intention to use mobile learning: A hybrid SEM-Neural Networks approach," *Comput Human Behav*, vol. 36, pp. 198–213, 2014.

[23] W. Paireekreng, K. Rapeepisarn, and K. W. Wong, "Time-based personalised mobile game downloading," *Transactions on Edutainment II*, pp. 59–69, 2009.

[24] S. S. Keerthi, S. K. Shevade, C. Bhattacharyya, and K. R. K. Murthy, "Improvements to Platt's SMO algorithm for SVM classifier design," *Neural Comput*, vol. 13, no. 3, pp. 637–649, 2001.

[25] I. H. Sarker, Y. B. Abushark, and A. I. Khan, "Contextpca: Predicting context-aware smartphone apps usage based on machine learning techniques," *Symmetry (Basel)*, vol. 12, no. 4, p. 499, 2020.

[26] J. Hong, E.-H. Suh, J. Kim, and S. Kim, "Context-aware system for proactive personalized service based on context history," *Expert Syst Appl*, vol. 36, no. 4, pp. 7448–7457, 2009.

[27] I. H. Sarker, H. Alqahtani, F. Alsolami, A. I. Khan, Y. B. Abushark, and M. K. Siddiqui, "Context pre-modeling: an empirical analysis for classification based user-centric context-aware predictive modeling," *J Big Data*, vol. 7, no. 1, pp. 1–23, 2020.

[28] L. D. Turner, S. M. Allen, and R. M. Whitaker, "Interruptibility prediction for ubiquitous systems: conventions and new directions from a growing field," in *Proceedings of the 2015 ACM International Joint Conference on Pervasive and Ubiquitous Computing*, 2015, pp. 801–812.

[29] V. Pejovic and M. Musolesi, "InterruptMe: designing intelligent prompting mechanisms for pervasive applications," in *Proceedings of the 2014 ACM International Joint Conference on Pervasive and Ubiquitous Computing*, 2014, pp. 897–908.

[30] A. Y.-L. Chong, "Predicting m-commerce adoption determinants: A neural network approach," *Expert Syst Appl*, vol. 40, no. 2, pp. 523–530, 2013.
[31] A. Dekel, D. Nacht, and S. Kirkpatrick, "Minimizing mobile phone disruption via smart profile management," in *Proceedings of the 11th International Conference on Human-Computer Interaction with Mobile Devices and Services*, 2009, pp. 1–5.
[32] H. Zhu, E. Chen, H. Xiong, H. Cao, and J. Tian, "Mobile app classification with enriched contextual information," *IEEE Trans Mob Comput*, vol. 13, no. 7, pp. 1550–1563, 2013.
[33] S. Hayat, F. Rehman, N. Riaz, H. Sharif, S. Irshad, and S. Shareef, "Using machine learning algorithms to detect dysplasia in Barrett's esophagus," in *2022 2nd International Conference on Digital Futures and Transformative Technologies (ICoDT2)*, IEEE, 2022, pp. 1–6.
[34] H. Sharif, F. Rehman, and A. Rida, "Deep learning: Convolutional neural networks for medical image analysis-a quick review," in *2022 2nd International Conference on Digital Futures and Transformative Technologies (ICoDT2)*, IEEE, 2022, pp. 1–4.
[35] S. Sun, C. Luo, and J. Chen, "A review of natural language processing techniques for opinion mining systems," *Information Fusion*, vol. 36, pp. 10–25, 2017.
[36] I. H. Sarker, M. M. Hoque, M. K. Uddin, and T. Alsanoosy, "Mobile data science and intelligent apps: concepts, ai-based modeling and research directions," *Mobile Networks and Applications*, vol. 26, no. 1, pp. 285–303, 2021.
[37] I. H. Sarker and A. S. M. Kayes, "ABC-RuleMiner: User behavioral rule-based machine learning method for context-aware intelligent services," *Journal of Network and Computer Applications*, vol. 168, no. 1, p. 102762, 2020.
[38] J. Lu, D. Wu, M. Mao, W. Wang, and G. Zhang, "Recommender system application developments: a survey," *Decis Support Syst*, vol. 74, no.1, pp. 12–32, 2015.
[39] E. Marine-Roig, E. Martin-Fuentes, and N. Daries-Ramon, "User-generated social media events in tourism," *Sustainability*, vol. 9, no. 12, p. 2250, 2017.
[40] K. Lee, *The World Health Organization (WHO)*. Routledge, 2008.
[41] I. H. Sarker, "Context-aware rule learning from smartphone data: survey, challenges and future directions," *J Big Data*, vol. 6, no. 1, pp. 1–25, 2019.
[42] I. H. Sarker, A. Colman, J. Han, A. I. Khan, Y. B. Abushark, and K. Salah, "Behavdt: a behavioral decision tree learning to build user-centric context-aware predictive model," *Mobile Networks and Applications*, vol. 25, no. 1, pp. 1151–1161, 2020.
[43] A. Maedche and S. Staab, "Ontology learning for the semantic web," *IEEE Intell Syst*, vol. 16, no. 2, pp. 72–79, 2001.

CHAPTER 5

Study on Routing Protocols and Ethical Machine Learning in Internet of Things Networks

Aziz Ullah, Abdul Siddique, Khalid Mahmood, Ahthasham, Mehak Mushtaq Malik, Abdullah Akbar, Mariya Ouaissa, and Mariyam Ouaissa

5.1 INTRODUCTION

The Internet of Things (IoT) has revolutionized the way we interact with and connect devices in various domains, including smart homes, industrial automation, healthcare, transportation, and agriculture. IoT networks consist of numerous devices, such as sensors, actuators, and smart objects, which generate and exchange vast amounts of data. This data is often transmitted over wireless communication channels and needs to be efficiently routed from source to destination [1]. As we embark on a study delving into Routing Protocols and Ethical Machine Learning in Internet of Things Networks, we aim to unravel the technical intricacies while keeping a mindful eye on the ethical dimensions that shape the very essence of this transformative technological landscape.

Traditional routing techniques designed for the Internet are not always suitable for IoT networks due to several inherent characteristics and challenges. IoT devices are typically resource-constrained, with limited processing power, memory, and battery life. Moreover, IoT deployments often involve heterogeneous networks, combining various wireless technologies, such as Wi-Fi, Bluetooth, ZigBee, and cellular networks. Mobility, scalability, security, and privacy are also critical concerns in IoT routing and ethical machine learning in this context [2].

DOI: 10.1201/9781032648323-5

Efficient and reliable routing is crucial to ensure timely delivery of data, optimal utilization of network resources, and overall network performance. Therefore, there is a need for specialized routing protocols, algorithms, and architectures that address the unique requirements and challenges of IoT environments. Research in this field aims to develop innovative solutions to optimize routing performance, reduce energy consumption, enhance security, and accommodate the dynamic nature of IoT networks [3].

Before we begin the survey of IoT routing, the fundamentals are explained in the following sections.

5.2 IOT-ENABLED ROUTING

Routing plays a critical role in IoT communication by determining the path through which data travels from source devices to destination devices. It involves making decisions on selecting the optimal routes, considering factors such as network topology, available resources, and data delivery requirements. Efficient routing protocols ensure timely and reliable data transmission, reducing latency, energy consumption, and network congestion.

The motivation for IoT routing research stems from the unique challenges posed by IoT networks. These challenges include the large-scale deployment of resource-constrained devices, the heterogeneity of device capabilities, dynamic network topologies, and the need for security and privacy.

Given the rapid growth of IoT deployments and the increasing demand for reliable and scalable communication, research in IoT routing is crucial. It aims to develop novel routing strategies, protocols, and algorithms that can meet the specific requirements of IoT networks and enable seamless connectivity and communication among diverse IoT devices. The subsequent sections of this chapter will delve into the fundamental concepts, challenges, existing routing protocols, and future research directions in IoT routing to address these pressing needs.

IoT has emerged as a transformative paradigm, connecting a vast array of devices and enabling seamless communication between them. It encompasses various smart devices such as sensors, actuators, and embedded systems, which interact and exchange data in a networked environment.

The significance of IoT lies in its ability to revolutionize multiple domains, including healthcare, transportation, manufacturing, and agriculture. IoT systems provide real-time monitoring, automation, and data-driven decision-making, leading to improved efficiency, productivity, and convenience. However, to enable seamless communication and efficient data exchange, effective routing strategies are essential.

Overview of IoT network architecture: IoT networks typically consist of a vast number of interconnected devices that form a complex network topology. These devices can be organized in various architectures such as star, mesh, or hybrid topologies. The architecture may include edge devices, gateways, cloud infrastructure, and backend servers. Understanding the structure and components of IoT network architecture is crucial for designing efficient routing mechanisms.

Characteristics of IoT traffic and communication patterns: IoT devices generate diverse types of traffic, including periodic sensor readings, event-driven alerts, and command and control messages. The traffic patterns in IoT networks can be sporadic, busty, or time-sensitive, depending on the application requirements. IoT communication often involves a large number of devices simultaneously transmitting data, requiring efficient handling of concurrent traffic flows and congestion control [4]. Routing algorithms must consider the energy consumption of devices and minimize overhead to extend the battery life of resource-constrained devices [5].

Importance of energy efficiency and scalability in IoT routing: IoT devices are often resource-constrained, operating on limited battery power or low-energy harvesting sources. Energy efficiency is a critical consideration in IoT routing to prolong the battery life of devices. Privacy concerns also arise due to the sensitive nature of IoT data, and routing protocols need to address privacy requirements while ensuring efficient data delivery [6].

Scalability is another crucial aspect of IoT routing. With the exponential growth of IoT devices, routing protocols must handle the increasing network size efficiently. Scalable routing mechanisms should be capable of managing the dynamic addition, removal, and mobility of devices, while ensuring low control overhead and minimal routing table sizes [7].

5.3 ROUTING PROTOCOLS FOR IOT

5.3.1 IPv6 Routing Protocol for Low- Power and Lossy Networks (RPL)

RPL is a widely used routing protocol specifically designed for low-power and lossy networks (LLNs) in IoT [8]. It utilizes the IPv6 addressing scheme and supports efficient routing in resource-constrained environments. RPL uses a Directed Acyclic Graph (DAG) structure to establish routes and incorporates features such as route optimization, traffic engineering, and support for multihop communication [7].

5.3.2 Ad Hoc On-Demand Distance Vector (AODV)

AODV is an on-demand routing protocol commonly used in ad hoc networks, including certain IoT scenarios [9]. It establishes routes dynamically as needed, reducing overhead and adapting to changing network conditions. AODV employs a distance vector algorithm to determine the shortest path between devices. It is suitable for IoT applications with intermittent connectivity or dynamic network topologies [10].

5.3.3 Constrained Application Protocol (CoAP)

CoAP is an application-layer protocol designed for IoT devices with constrained resources. It enables lightweight communication between devices and is often used in conjunction with IP networks. CoAP utilizes a client–server model and supports request–response interactions, multicast communication, and resource discovery [11]. It is well-suited for IoT applications where low overhead and energy efficiency are paramount.

5.3.4 Message Queuing Telemetry Transport (MQTT)

MQTT is a publish-subscribe messaging protocol commonly used in IoT deployments [12]. It follows a lightweight publish-subscribe model, enabling efficient data distribution among devices. MQTT uses a broker-based architecture, where devices publish messages to topics, and subscribers receive the messages of interest. It is suitable for IoT applications that require real-time data updates and efficient network utilization.

5.4 ETHICAL MACHINE LEARNING IN IOT NETWORKS

Imagine we have a personal assistant that helps us make decisions based on the information it gathers about us and our surroundings. Ethical machine learning is like training this assistant to not only be smart but also fair, unbiased, and respectful in its decision-making process. IoT networks is like a bustling community of interconnected devices—smart home gadgets, industrial sensors, healthcare monitors, and more—all communicating and sharing information with each other [13]. It is like a digital ecosystem where devices work together to make our lives easier. Now, picture all these devices constantly generating and exchanging data. The challenge arises when we introduce machine learning into this mix. Routing protocols need to optimize energy consumption by minimizing data transmission distances, using sleep/wake-up strategies, and leveraging energy-aware routing algorithms [14]. The algorithms that make decisions based on this data need to be not just accurate but also mindful of ethical considerations [15].

Ethical concerns in this context revolve around ensuring fairness, transparency, and respect for privacy [16]. It is about making sure that these machine learning algorithms do not unintentionally favor one group over another, that they operate openly so we can understand how decisions are made, and that they handle our personal data with the utmost care. Ethical machine learning is not just the responsibility of developers and engineers. It involves everyone in the ecosystem, from policymakers to end-users [17]. It is a collective effort to ensure that as technology advances, it does so in a way that respects our values and preserves our humanity. Addressing these challenges requires innovative routing algorithms and protocols that can optimize network performance, accommodate device constraints, and ensure data integrity and confidentiality [18]. The proliferation of IoT devices has resulted in an exponential growth of data traffic and raised the need for efficient data routing mechanisms [19].

5.5 CHALLENGES

In the world of smart IoT devices like trackers and smart devices, there are two big challenges. First, we need these devices to talk to each other using as little battery power as possible, so they do not die on us. It is like making sure your gadgets are energy-efficient superheroes. Second, when these devices learn about us, it is important they treat everyone fairly and keep our secrets. It is a bit like having a super-smart friend who knows our habits but never spills the beans and always treats everyone equally. Thus, the challenge is to make sure our gadgets are both smart and respectful.

5.5.1 Resource-Constrained Nature of IoT Devices

IoT devices often have limited processing power, memory, and energy resources. These constraints pose challenges in designing routing protocols that are lightweight and efficient in terms of resource utilization. Algorithms and data structures for routing protocols must be optimized to use less memory and CPU. They ought to keep the size of routing tables kept on devices to a minimum and minimize control packet overhead. Reducing data transmission and reception can also help save electricity. Protocols can make use of techniques like dynamic duty-cycling of radios based on flow patterns, employing effective protocol types, and compressing routing tables.

5.5.2 Heterogeneity and Mobility of IoT Devices

IoT networks consist of devices with diverse capabilities, communication interfaces, and mobility patterns. Heterogeneity introduces challenges in establishing communication and selecting appropriate routing paths based on the device characteristics. Mobility of IoT devices adds further complexity as routing protocols need to adapt to changing network topologies due to device movement [20]. Routing protocols need to be able to identify device characteristics and choose the best routes taking into account the processing capacity, mobility patterns, and available interfaces of each device.

5.5.3 Scalability and Dynamic Topology Management

IoT networks can scale to include a massive number of devices, leading to challenges in managing routing tables and controlling network overhead. Routing protocols must handle the dynamic addition and removal of devices while maintaining efficient routing paths [21]. They should be capable of discovering and adapting to changes in network topologies in real-time to ensure seamless communication. Decentralized methods are required for topology management in routing protocols. They have to dynamically update pathways in reaction to changes and efficiently disseminate routing information among a large number of devices.

5.5.4 Ethical Machine Learning Security and Privacy Concerns

IoT networks are vulnerable to various security threats, including unauthorized access, data breaches, and malicious attacks [22]. This is where we have to reconsider the data we are given for IoT routing. Routing protocols must incorporate robust security mechanisms to ensure secure data transmission, authentication, and confidentiality. Routing protocols should use security features like encryption, integrity checks, and access control to guard against spoofing, tampering, and eavesdropping.

5.5.5 Standards for Quality of Service (QoS)

Because different IoT applications have different and strict requirements for quality of service, it might be difficult to ensure reliable data transport within the given timeliness, bandwidth, and reliability restrictions. Mission-critical data applications in fields like automated manufacturing and healthcare surveillance have minimal latency requirements

[23]. Routing protocols must priorities such time-sensitive traffic and allow QoS-aware path selection. They have to meet the varying throughput and reliability requirements of various IoT use cases while handling concurrent transmissions in an effective manner [24]. The efficiency and functionality of IoT systems may suffer if QoS criteria are not met.

5.5.6 Fault Tolerance

Node and link failures are common in IoT networks, which are made up of a huge number of inexpensive devices placed in a variety of locations. Topological alterations and disturbances can be brought on by device malfunctions, low battery levels, connectivity issues, etc. [25]. It can be difficult for routing protocols to identify these problems quickly and isolate malfunctioning parts in order to keep the network connected [26]. To prevent disruptions, they have to allow for fast rerouting of traffic through other routes. To ensure the dependability of operations in IoT applications, self-healing methods are also required to automatically repair paths around problems without human intervention [27].

5.5.7 Limitations in Energy Harvesting

Due to limitations on battery replacement, a large number of IoT devices in isolated or difficult-to-reach places depend on gathering energy from external sources such as solar, thermal, kinetic, etc. [28]. Scavenged energy is unpredictable and sporadic, which makes it difficult for routing protocols to adjust to changes in power availability [29]. In order to maximize energy utilization based on available reserves and harvesting rates, they require effective methods [29]. Adaptive duty-cycling and traffic load-balancing strategies might help autonomous energy harvesting IoT devices to maintain dependable connectivity and maximize operational lifetimes despite fluctuating energy restrictions [30].

5.6 SURVEY ANALYSIS

To gain insights into the current trends and challenges in ethical machine learning in IoT routing, a survey was conducted among professionals and researchers in the field. The survey aimed to understand their perspectives on various aspects of IoT routing protocols, algorithms, and challenges. There is a need to be careful about how we use machine learning in our smart gadgets, especially when it comes to deciding how they talk to each other. People are worried about keeping info private, making sure these algorithms are transparent and fair, and avoiding any unintentional biases. The survey received a total of 150 responses, and the data was analyzed to derive meaningful insights.

5.6.1 Participant Profile

The participants in the survey consisted of professionals and researchers from diverse backgrounds, including academia, industry, and IoT-related organizations. The majority of respondents (65%) had more than three years of experience working with IoT networks and routing. The participants represented a wide range of domains, including smart cities, healthcare, manufacturing, and agriculture. Participants' varied experiences and areas of expertise allowed for well-rounded insights on the application and technical elements of IoT routing. Academics presented perspectives focused on research, while industry experts shared ideas on practical implementations.

5.6.2 IoT Routing Protocols Usage

The survey examined the usage of different IoT routing protocols among the participants. The results showed that RPL was the most widely adopted routing protocol, with 45% of the respondents indicating its usage in their IoT deployments. AODV and CoAP were also commonly used, with adoption rates of 30% and 25%, respectively. MQTT, being a popular messaging protocol, was utilized by 35% of the participants for specific IoT communication requirements. RPL, AODV, and CoAP are widely used, indicating that they successfully handle common IoT routing scenarios. But no single protocol can satisfy every demand, hence continual research into specific solutions for particular application domains is required.

5.6.3 Factors Influencing Protocol Selection

Participants were asked about the factors influencing their choice of IoT routing protocols. The analysis revealed that energy efficiency was the most important factor (cited by 60% of the respondents). Scalability and support for heterogeneous networks were also significant factors (mentioned by 40% and 35% of the participants, respectively). Security and reliability were considered important by 25% and 20% of the respondents, respectively. Energy efficiency will gain relevance as IoT deployments increase to include more devices with limited resources. Reliability, heterogeneity, and scalability concerns will guarantee that protocols continue to satisfy the changing needs of extensive IoT networks.

5.6.4 Key Challenges in IoT Routing

The survey investigated the key challenges faced by professionals and researchers in IoT routing. The results showed that the resource-constrained nature of IoT devices was the most significant challenge, with 55% of the participants highlighting it as a major concern. Heterogeneity and mobility of devices were also recognized as significant challenges (mentioned by 40% and 35% of the respondents, respectively). Other challenges identified included scalability (25%) and security/privacy (20%). Developing novel strategies is necessary to overcome the limitations low-cost devices present. More protocol improvements are needed in the areas of handling mobility and connectivity to support a variety of IoT deployment scenarios.

5.6.5 Future Research Directions

Participants were asked to suggest potential research directions for improving ethical machine learning perspective in IoT routing. The analysis of their responses revealed several interesting areas for future exploration. Integration of IoT with 5G networks (45%) emerged as a prominent research direction, indicating the potential of leveraging

5G capabilities to enhance IoT routing performance [31]. Blockchain-based routing solutions (30%) and protocol enhancements and standardization efforts (25%) were also identified as areas for further research and development.

Overall, the survey analysis provided valuable insights into the current practices, challenges, and future research directions in IoT routing. The findings indicate the importance of energy efficiency, scalability, and addressing resource constraints in IoT routing protocols. They also highlight the need for further exploration of emerging technologies like 5G and blockchain in the context of IoT routing [32]. Constant progress in distributed ledger and networking technology offers chances to create new approaches to open problems in the IoT routing space.

5.7 FUTURE WORK AND CONCLUSION

IoT routing plays a crucial role in enabling seamless communication and efficient data exchange in the vast and diverse landscape of the IoT. In this chapter, we provided an introduction to IoT routing, discussing the role of ethical machine learning in IoT routing networks and emphasizing its significance in facilitating effective communication among IoT devices [33].

We explored the fundamental concepts, challenges, and existing routing protocols in IoT networks. The resource-constrained nature of IoT devices, heterogeneity and mobility, scalability and dynamic topology management, as well as security and privacy concerns, pose significant challenges in designing efficient routing solutions.

The chapter presented an overview of prominent IoT routing protocols, such as RPL, AODV, CoAP, and MQTT, highlighting their strengths and limitations. It is evident that no single protocol fits all IoT scenarios, and the choice of routing solution depends on specific application requirements.

While significant progress has been made in IoT routing research, there are still open challenges to address. Integrating IoT with 5G networks, exploring blockchain-based routing solutions, enhancing protocol standardization, managing transportation, and conducting real-world evaluations are areas that demand further exploration [34].

REFERENCES

[1] Safara, F., Souri, A., Baker, T., Al Ridhawi, I., & Aloqaily, M. (2020). PriNergy: A priority-based energy-efficient routing method for IoT systems. Journal of Supercomputing, 76(11), 8609–8626.

[2] Park, S., Crespi, N., Park, H., & Kim, S.H. (2014, March). IoT routing architecture with autonomous systems of things. In *2014 IEEE World Forum on Internet of Things (WF-IoT)* (pp. 442–445). IEEE.

[3] Muhammad, M., Sajid, A., & Arshad, H. (2023). Layer-Based Security Threats in IoT Networks. 10.1201/9781003404361-6.

[4] Tahaei, H., Afifi, F., Asemi, A., Zaki, F. & Anuar, N.B. (2020). The rise of traffic classification in IoT networks: A survey. Journal of Network and Computer Applications, 154, 102538.

[5] Dhumane, A., and Prasad, R. (2015). Routing challenges in internet of things. CSI Communications, 19–20.

[6] Hameed, S., Khan, F.I. and Hameed, B. (2019). Understanding security requirements and challenges in Internet of Things (IoT): A review. Journal of Computer Networks and Communications, 1, pp.1–14.

[7] Xin, H.M. & Yang, K. (2015). Routing protocols analysis for Internet of Things. In *2015 2nd international conference on information science and control engineering* 2015 Apr 24 (pp. 447–450). IEEE.

[8] David, A., Gutierrez, J. & Ray, S.K. (2016, August). A lightweight trust design for IoT routing. In *2016 IEEE 14th Intl Conf on Dependable, Autonomic and Secure Computing, 14th Intl Conf on Pervasive Intelligence and Computing, 2nd Intl Conf on Big Data Intelligence and Computing and Cyber Science and Technology Congress (DASC/PiCom/DataCom/CyberSciTech)* (pp. 552–557). IEEE.

[9] Ullah Khan, I., Mansoor Qureshi, I., Aziz, M.A., Cheema, T-W.A. & Shah, S.B.H. (2020). Smart iot control-based nature inspired energy efficient routing protocol for flying ad hoc network (fanet). IEEE Access, 8:56371–56378.

[10] Aleyadeh, S. & Taha, A.E.M. (2018, May). An IoT-Based architecture for waste management. In *2018 IEEE International Conference on Communications Workshops (ICC Workshops)*, pp. 1. IEEE.

[11] Ullah Khan, I., Yaseen Ayub, M., Abdollahi, A., and Dutta, A. (2023). A hybrid deep learning model-based intrusion detection system for emergency planning using IoT-network. In *2023 International Conference on Information and Communication Technologies for Disaster Management (ICT-DM)* (pp. 1–5).

[12] Ullah Khan, I., Abdollahi, A., Jamil, A., Baig, B., Adnan Aziz, M., and Subhan, F. (2022). A novel design of fanet routing protocol aided 5g communication using IoT. Journal of Mobile Multimedia, (1), pp. 1333–1354, 2022

[13] Ullah Khan, I., Alturki, R., Alyamani, H.J., Ikram, M.A., Aziz, M.A., Hoang, V.T., & Cheema, T.A. (2021). Rssi-controlled long-range communication in secured IoT-enabled unmanned aerial vehicles. Mobile Information Systems, 2021, pp. 1–11.

[14] Ullah Khan, I., Zukhraf, S.Z.N., Abdollahi, A., Imran, S.A., Mansoor Qureshi, I., Aziz, M.A., and Shah, S.B.H. (2020). Reinforce based optimization in wireless communication technologies and routing techniques using internet of flying vehicles. In *Proceedings of the 4th International Conference on Future Networks and Distributed Systems*, pp. 1–6.

[15] Rehman, H.U., Asif, M. & Ahmad, M. (2017, December). Future applications and research challenges of IOT. In *2017 International conference on information and communication technologies (ICICT)* (pp. 68–74). IEEE.

[16] Ullah Khan, I., Hassan, M.A., Fayaz, M., Gwak, J., and Aziz, M.A. (2022). Improved sequencing heuristic dsdv protocol using nomadic mobility model for fanets. Computers, Materials & Continua, 70(2), pp. 3653–3666, 202

[17] Abdollahi, A. & Fathi. M. (2020). An intrusion detection system on ping of death attacks in IoT networks. Wireless Personal Communications, 112:2057–2070.

[18] Ullah Khan, I., Shah, S.B.H., Wang, L., Aziz, M.A., Stephan, T., & Kumar, N. (2021). Routing protocols & unmanned aerial vehicles autonomous localization in flying networks. *International Journal of Communication Systems*, (1), p. e4885.

[19] Ullah Khan, I. , Ouaissa, M., Houda, Z.A.E., & Ijaz, M.H. (2024). *Cyber Security for Next-Generation Computing Technologies*. CRC Press.

[20] Felzmann, H., Villaronga, E.F., Lutz, C. & Tam`o-Larrieux, A. (2019). Transparency you can trust: Transparency requirements for artificial intelligence between legal norms and contextual concerns. Big Data & Society, 6(1), p. 2053951719860542.

[21] Singh, G., Kumar, N., & Verma, A.K. (2014). Antalg: An innovative aco based routing algorithm for manets. Journal of Network and Computer Applications, 45, pp. 151–167.

[22] Tennant, P.W.G., Murray, E.J., Arnold, K.F., Berrie, L., Fox, M.P., Gadd, S.C., Harrison, W.J., Keeble, C., Ranker, L.R., Textor, J. et al. (2021). Use of directed acyclic graphs (dags) to identify confounders in applied health research: review and recommendations. International Journal of Epidemiology, 50(2), pp. 620–632.

[23] Zanella, A., Bui, N., Castellani, A., Vangelista, L. & Zorzi, M. (2014). Internet of things for smart cities. IEEE Internet of Things Journal, 1(1), pp.22–32.

[24] Yang, P., Xiao, Y., Xiao, M. & Li, S., (2019). 6G wireless communications: Vision and potential techniques. IEEE Network, 33(4), pp.70–75.

[25] De Souza, L.M.S., Vogt, H. & Beigl, M. (2007). A survey on fault tolerance in wireless sensor networks. Interner Bericht. Fakultät für Informatik, Universität Karlsruhe.

[26] Willig, A. (2008). Recent and emerging topics in wireless industrial communications: A selection. IEEE Transactions on Industrial Informatics, 4(2), pp.102–124.

[27] Willig, A., Matheus, K. & Wolisz, A. (2005). Wireless technology in industrial networks. Proceedings of the IEEE, 93(6), pp.1130–1151.

[28] Zhang, Q., Wang, Y. & Kim, E.S., (2015). Electromagnetic energy harvester with flexible coils and magnetic spring for 1–10 Hz resonance. Journal of Microelectromechanical Systems, 24(4), pp.1193–1206.

[29] Roundy, S. & Wright, P. K. (2004). A piezoelectric vibration based generator for wireless electronics. Smart Materials and Structures, vol. 13, no. 5, pp. 1131–1142.

[30] Ko, J., Lu, C., Srivastava, M.B., Stankovic, J.A., Terzis, A. & Welsh, M., (2010). Wireless sensor networks for healthcare. Proceedings of the IEEE, 98(11), pp.1947–1960.

[31] Yi, C., Abraham, J., Afanasyev, A., Wang, L., Zhang, B., & Zhang, L. (2014). On the role of routing in named data networking. In *Proceedings of the 1st ACM Conference on Information-Centric Networking* (pp. 27–36).

[32] Sanjuan, E. B., Cardiel, I. A., Cerrada, J. A., & Cerrada, C. (2020). Message queuing telemetry transport (mqtt) security: A cryptographic smart card approach. IEEE Access, 8, pp. 115051–115062.

[33] Oosman, B. S., & Dudhe, R. (2021). Review on the ethical and legal challenges with IoT. In *2021 International Conference on Computational Intelligence and Knowledge Economy (ICCIKE)* (pp. 529–534), IEEE.

[34] Usama, M., Ullah, U. & Sajid, A. (2024). Cyber Attacks Against Intelligent Transportation Systems. In *Cyber Security for Next-Generation Computing Technologies* (pp. 190–230). CRC Press. 10.1201/9781003404361-11.

CHAPTER 6

Reinventing Artificial Intelligence and Blockchain for Preserving Medical Data

Bhupinder Singh and Christian Kaunert

6.1 INTRODUCTION

Ensuring secure data sharing and collaboration is of paramount importance in healthcare analytics. The intricate nature of healthcare data demands a meticulous approach to protect patient privacy, comply with regulatory requirements, and harness the potential of data-driven insights. Employing advanced encryption techniques and access controls, healthcare organizations can establish a robust framework that safeguards sensitive information during both transmission and storage. Also, de-identification practices play a pivotal role in shielding patient identities while allowing for meaningful analysis. By implementing data use agreements, obtaining informed consent, and upholding ethical considerations, healthcare professionals can strike a balance between analytical exploration and patient confidentiality [1]. Leveraging blockchain technology can provide transparent and tamper-proof data sharing, enhancing trust among collaborating parties. Regular security audits, role-based access controls, and comprehensive breach response plans further fortify the data-sharing ecosystem. As healthcare stakeholders forge collaborative partnerships, fostering a culture of awareness and training ensures that data sharing remains a dynamic force for innovation while upholding the highest standards of security and patient well-being [2].

The integration of artificial intelligence (AI) and blockchain technology holds significant promise for preserving privacy and enhancing the security of medical data. AI, with its capacity for data analysis and pattern recognition, and blockchain, with its

DOI: 10.1201/9781032648323-6

decentralized and tamper-resistant ledger, create a synergistic partnership that addresses some of the most critical challenges in healthcare data management.

Using methods like federated learning, AI can significantly improve the privacy of medical data. This method eliminates the need to centralize sensitive data by training AI models locally on dispersed data sources, like clinics or hospitals. The models ensure that each patient record stays on its own property while learning from the data. This allows the AI model to learn from a variety of sources while maintaining patient privacy because the data never leaves its original place. Blockchain technology has the potential to offer a transparent and safe framework for the management of medical data. Sensitive data, like medical histories and patient records, can be safely kept on a blockchain, where every modification or transaction is logged as an unchangeable, time-stamped entry. It guards against unwanted alteration and guarantees data integrity. With smart contracts, patients can give researchers and healthcare professionals access to particular data while maintaining control over their data, allowing them to withdraw access at any point in time [3].

Through encryption and decentralized access management, the combination of AI and blockchain can further improve the security of medical data. Patient data can be encrypted using AI algorithms so that only authorized persons with the right decryption keys can access it. By implementing access control methods on the blockchain, patients may securely and auditably govern permissions for their data. Anomaly detection driven by AI can also keep an eye on blockchain transactions for any odd activity and notify managers of possible security breaches. The medical data ecosystem's overall security is improved by this proactive approach. Advanced data exchange and interoperability are also made possible by the combination of blockchain technology and AI. The collaborative research endeavors are facilitated by the secure flow of data among healthcare practitioners, researchers, and patients across various systems and platforms, all while maintaining data privacy and permission. The amalgamation of blockchain technology and AI presents a potent resolution for safeguarding medical data security and maintaining privacy [4]. The integration of blockchain technology's decentralized, tamper-resistant architecture with AI's analytical powers can enable healthcare organizations to build a strong, patient-focused data management ecosystem that puts privacy, security, and interoperability first. This innovative synergy has the potential to revolutionize the way medical data is handled, shared, and utilized, ushering in a new era of patient empowerment and data-driven healthcare advancements.

6.1.1 Relevance of Chapter

Secure data sharing and collaboration in healthcare analytics represents a vital equilibrium between leveraging data-driven insights and safeguarding patient privacy. Achieving this delicate balance involves employing advanced technological safeguards, adhering to legal regulations such as HIPAA, and implementing stringent ethical considerations. Establishing transparency and trust among stakeholders is facilitated through the implementation of encryption, secure data-sharing platforms, role-based access controls, regular security audits, and blockchain technology. Clear data use agreements and informed consent mechanisms further fortify this foundation. Additionally, role-based access controls, regular security audits, and blockchain technology bolster data integrity and mitigate risks [5]. By meticulously integrating these measures, healthcare analytics can thrive as a collaborative endeavor that not only generates transformative insights but also upholds the paramount importance of patient confidentiality and data security.

There are various international legal frameworks and guidelines that offer valuable insights into the ethical and legal considerations surrounding their collection and use. Universal instruments such as the Universal Declaration on Bioethics and Human Rights underscore the importance of respecting individual autonomy, privacy, and the protection of personal data in medical research and healthcare activities. The World Health Organization's Ethical Guidelines for Health-related Research Involving Humans emphasize informed consent, benefit sharing, and the safeguarding of privacy and confidentiality when dealing with medical data. The European General Data Protection Regulation (GDPR) establishes a robust standard for data protection and privacy within the European Union, affecting the handling of medical data within its jurisdiction. Furthermore, the Council of Europe's Convention for the Protection of Individuals with regard to Automatic Processing of Personal Data addresses principles applicable to personal data, including medical information, in automated processing [6]. As the landscape of big data and medical data continues to evolve, international organizations and standards bodies, such as the International Organization for Standardization (ISO), contribute with guidelines related to data security, privacy, and health informatics. Although a comprehensive international law specifically tailored to big data and medical data is absent, these existing frameworks collectively highlight the importance of ethical data practices, informed consent, individual rights, and data security when dealing with the intricacies of medical information within the context of evolving data analytics paradigms.

In the Indian legal landscape, the intersection of big data and medical data is a complex and evolving realm that requires a delicate balance between leveraging the immense potential of data-driven insights and safeguarding the confidentiality and privacy of sensitive healthcare information. While India lacks a dedicated comprehensive legislation solely focused on big data in healthcare, several key legal instruments play a pivotal role in governing the collection, storage, and use of medical data within the context of big data analytics. The Information Technology Act, 2000, and its subsequent rules establish the foundational principles for electronic transactions and data protection, providing a framework for safeguarding medical data during electronic exchange and storage [7]. Additionally, the forthcoming Personal Data Protection Bill, 2019 seeks to establish robust data protection norms, including healthcare-related data, by introducing principles such as informed consent, data localization, and individual rights, which are especially pertinent to big data utilization in the healthcare domain.

The Medical Council of India's ethical regulations and guidelines underscore the importance of patient confidentiality, informed consent, and ethical data practices for medical professionals, laying the groundwork for responsible handling of medical data in the era of big data analytics. The evolving landscape of healthcare, combined with rapid technological advancements, underscores the need for continuous legal adaptations and refinements to ensure that India's legal framework effectively addresses the ethical, privacy, and security considerations inherent in the intersection of big data and medical data. As India embarks on its journey towards a comprehensive data protection regime, the nuanced interplay between medical data and big data continues to be a subject of legal discourse, necessitating vigilance in upholding patient rights, data integrity, and responsible data governance [8].

6.1.2 Objectives of the Chapter

The chapter outlines the objectives that form the framework for researching the integration of blockchain technology and AI in the healthcare sector, which are as follows:

- list and evaluate the current issues with data management in healthcare, such as interoperability, security, and data preservation to give a thorough grasp of the problems the healthcare industry needs to solve.
- investigate the possibility of using blockchain technology to address the difficulties associated with managing patient data. This entails comprehending the ways in which blockchain might improve data security, integrity, and privacy as well as its function in facilitating data sharing and interoperability within the healthcare industry.
- analyze the function of AI in health analytics, emphasizing its use in decision assistance, data analysis, and personalized treatment to draw attention to the importance of AI in healthcare and how it may enhance patient outcomes.
- examine the ways in which blockchain and AI integration can offer novel approaches to healthcare data analytics and management. This entails examining use cases and possible advantages of merging these technologies as well as evaluating how they affect the provision of healthcare.
- inspect the legal ramifications of using blockchain technology in healthcare including patient data privacy, consent, and regulatory issues as well as offer legal opinions and suggestions to guarantee the ethical and responsible application of these technologies in healthcare [9].

6.1.3 Structure of Chapter

The chapter provides a comprehensive review and analysis on medical data and the applications of AI to preserve it. Section 6.2 specifies the factors associated with secure data sharing and collaboration in healthcare analytics. Section 6.3 explores the concept of sharing of healthcare data in multi-cloud. Section 6.4 discusses the vital equilibrium between leveraging data-driven insights and safeguarding patient privacy. Section

6.5 discusses IoT and blockchain technology in cutting-edge methods for data sharing and privacy. Section 6.6 covers proposed solutions regarding legal aspects. And, finally Section 6.7 provides a conclusion and future scope.

6.2 FACTORS ASSOCIATED WITH SECURE DATA SHARING AND COLLABORATION IN HEALTHCARE ANALYTICS

Secure data sharing and collaboration in healthcare analytics are essential components used to harness the power of data for informed decision-making and improved patient outcomes while maintaining patient privacy and data security. Guaranteeing that sensitive healthcare data is shared and analyzed in a secure and morally responsible manner requires a combination of technology advancements, regulatory frameworks, and best practices, as discussed in the following [10].

Access control and data encryption: Use strong encryption methods to safeguard data while it is being transferred and stored. It is imperative to implement access control measures to guarantee that data analysis and access are restricted to authorized individuals or entities.

Secure Data Sharing Platforms: To enable data sharing between authorized parties, make use of secure platforms that go by industry standards and laws, such as the Health Insurance Portability and Accountability Act (HIPAA) in the United States. End-to-end encryption, authentication, and audit trails should all be available on these platforms.

De-identification: To exclude personally identifiable information (PII) from patient data while maintaining the data's analytical usefulness, de-identify the data prior to sharing it. This helps protect patient privacy while enabling meaningful insights.

Data Use Agreements: Establish clear data use agreements that outline the purpose, scope, and responsibilities of each party involved in data sharing and collaboration. These agreements should detail data security measures, access controls, and limitations on data use.

Consent and Ethical Considerations: Obtain informed consent from patients for data sharing and analytics whenever applicable. Uphold ethical principles by ensuring transparency, respecting patient autonomy, and safeguarding vulnerable populations.

Blockchain Technology: Consider utilizing blockchain technology for secure and transparent data sharing. Blockchain decentralized nature and cryptographic techniques enhance data integrity and auditability.

Data Minimization: Share only the minimum necessary data required for collaborative analytics. Reducing data exposure limits potential risks and privacy breaches.

Role-Based Access: Implement role-based access control to ensure that individuals have access to data based on their specific roles and responsibilities. This minimizes the risk of unauthorized data access.

Regular Security Audits: Conduct regular security audits and assessments of data sharing platforms and analytical tools to identify vulnerabilities and ensure compliance with security standards.

Training and Awareness: Train healthcare professionals and data analysts on data security best practices, privacy regulations, and ethical considerations to foster a culture of responsible data sharing.

Third-Party Vetting: Before collaborating with external partners, vet their data security practices and compliance with regulatory requirements to ensure a high level of trust.

Data Breach Response Plan: Develop a comprehensive data breach response plan that outlines steps to take in case of a security incident. Legal requirements, risk mitigation techniques, and communication plans should all be part of this plan.

Healthcare companies and analytics teams can maintain patient privacy and adhere to legal and ethical requirements by using these tactics and creating a secure environment for data exchange and collaboration [11]. By doing this, the healthcare sector can protect sensitive data, uphold stakeholder and patient trust, and extract useful insights from data.

6.3 SHARING OF HEALTHCARE DATA IN MULTI-CLOUD

Modern healthcare data management gains a complex yet necessary dimension with the introduction of multi-clouds, or the sharing of healthcare data across numerous cloud computing platforms. Benefits from multi-cloud techniques include enhanced resource optimization, redundancy, and scalability.

A strategic and complex task, sharing healthcare data in multi-cloud systems has the potential to transform the healthcare industry by improving resource optimization, scalability, and cooperation. By distributing healthcare data over several cloud computing platforms, this strategy enables healthcare companies to take use of a wide range of technological capabilities and easily share vital information [12]. However, a careful strategy is necessary to ensure the safe and responsible sharing of this information due to the complex nature of healthcare data, which is defined by its sensitivity, regulatory requirements, and ethical issues. Strong security measures, including data encryption, strict access controls, and data masking to protect patient privacy and stop unwanted access, must be in place in order to do this. In order to stay out of legal hot water and protect patient rights, compliance with healthcare regulations is essential. Examples of these legislation are the General Data Protection Regulation (GDPR) and the Health Insurance Portability and Accountability Act (HIPAA). Healthcare data stays in the jurisdiction where it is governed by regulations thanks to the establishment of clear data residency policies. Secure application programming interfaces (APIs) combined with auditing tools that record every data contact strengthen accountability and transparency and enable a thorough grasp of data flows and usage [13]. Data loss prevention solutions add an additional layer of protection by monitoring and preventing unauthorized data movement, reducing the risk of inadvertent breaches.

In this intricate web of interconnected clouds, careful consideration must be given to vendor selection, evaluating the security practices and certifications of cloud service providers to ensure they align with the stringent demands of healthcare data management. Moreover, redundancy and disaster recovery strategies guarantee data availability in the event of service disruptions or failures. As the healthcare industry increasingly relies on data-driven insights for clinical decision-making and research, the sharing of healthcare data in multi-cloud environments offers an avenue for accelerated innovation [14]. By fostering secure data sharing and interoperability, multi-cloud approaches empower healthcare stakeholders to collaboratively harness the power of data, catalyzing advancements that hold the potential to improve patient care, drive medical research, and transform the healthcare ecosystem as a whole. However, sharing sensitive healthcare data in this distributed environment necessitates careful consideration of security, compliance, and data integrity. Implementing secure healthcare data sharing in multi-clouds involves several key considerations:

Data Encryption: Encrypting healthcare data before sharing it across multi-clouds ensures that even if unauthorized access occurs, the data remains unreadable. Utilizing strong encryption protocols and key management systems enhances data security.

Access Control: Implement robust access control mechanisms that restrict data access to authorized personnel only. Role-based access control (RBAC) ensures that users have appropriate permissions based on their roles and responsibilities.

Data segmentation: Apply the least privilege principle to limit access to distinct segments and segregate sensitive patient information from non-sensitive data. By doing this, exposure and possible breaches are reduced.

Compliance: Make sure that healthcare laws, like the General Data Protection Regulation (GDPR) in the European Union and the Health Insurance Portability and Accountability Act (HIPAA) in the United States, are followed. To prevent legal ramifications, multi-cloud environments need to comply with all applicable legislation.

Data residence: Take into account the restrictions pertaining to data residence and make sure that medical records stay in the jurisdiction where compliance rules are in effect. Processing and storing data should comply with local legal standards.

Data Masking: To preserve patient privacy, use methods like data masking prior to sharing data. These methods allow data to be used for analysis while preserving individual identities.

Secure APIs: Implement secure application programming interfaces (APIs) to facilitate data sharing between different clouds. Secure API design helps prevent unauthorized access and data leakage.

Audit Trails: Maintain comprehensive audit trails that log all data access and modifications. These logs enable tracking and accountability in case of any security incidents.

Data Loss Prevention (DLP): Deploy DLP solutions that monitor and prevent the unauthorized movement of sensitive data between clouds. DLP tools help enforce security policies and prevent data leaks.

Redundancy and Disaster Recovery: Ensure data redundancy and disaster recovery plans are in place to safeguard healthcare data against potential cloud service outages or failures.

Vendor Security: Evaluate the security practices and certifications of cloud service providers. Choose reputable vendors that adhere to stringent security standards.

The sharing of healthcare data in multi-cloud environments offers scalability and flexibility but requires meticulous planning to ensure the security, compliance, and integrity of patient information. By implementing encryption, access controls, data segmentation, and other security measures, healthcare organizations can confidently embrace the benefits of multi-cloud while safeguarding sensitive data and maintaining regulatory compliance.

6.4 VITAL EQUILIBRIUM BETWEEN LEVERAGING DATA-DRIVEN INSIGHTS AND SAFEGUARDING PATIENT PRIVACY

The significance of healthcare analytics and big data sharing is profound, revolutionizing the way healthcare is delivered, managed, and improved. These transformative practices are driving a paradigm shift in the healthcare industry, offering a multitude of benefits that ultimately enhance patient outcomes, operational efficiency, and public health at large [15].

The implication of healthcare analytics and big data sharing in the modern healthcare landscape cannot be overstated, as these practices stand as transformative pillars reshaping the way healthcare is understood, delivered, and improved. The ability to leverage large, intricate datasets to derive actionable insights that could transform patient care, operational performance, and the efficacy of the health system as a whole is what makes this importance possible. Healthcare administrators and professionals can make well-informed decisions that go beyond convention and intuition by utilizing advanced data analytics approaches, which improves patient safety and clinical outcomes. This importance is further enhanced by the combination of big data sharing which allow different datasets from different sources such as wearables, genomics research, and electronic health records to be combined [16]. This community data repository acts as a knowledge base, promoting an unprecedented level of comprehension of health trends, disease patterns, and treatment responses. From early disease detection and prevention strategies to the personalized tailoring of treatment plans, the utilization of big data sharing provides a panoramic view of healthcare that transcends individual experiences and contributes to the advancement of population health.

The healthcare analytics and big data sharing underpin the quest for innovation, fueling groundbreaking research endeavors, clinical trials, and medical discoveries. These practices expedite the identification of novel drug candidates, facilitate the evaluation of treatment efficacy, and empower researchers to explore uncharted avenues of medical science. In parallel, the optimization of resource allocation and healthcare operations emerges as a critical facet of their significance, leading to streamlined workflows, reduced costs, and heightened operational efficiencies that ultimately benefit both healthcare institutions and patients alike [17].

In the context of public health, healthcare analytics and big data sharing emerge as indispensable tools for timely response and containment of outbreaks, accurate prediction of disease trends, and evidence-based policymaking. The ability to swiftly analyze vast datasets enables health authorities to monitor the spread of infectious diseases, deploy resources to high-risk areas, and make data-driven recommendations that safeguard entire communities. However, this significance does not come without its responsibilities. The ethical handling of sensitive patient data and the assurance of data security and privacy are paramount. Thus, striking a balance between harnessing the potential of healthcare analytics and big data sharing while safeguarding individual rights and data integrity necessitates robust legal frameworks, stringent regulations, and continuous oversight [18]. The impact of healthcare analytics and big data sharing lies in their capacity to propel healthcare into a new era of precision, efficiency, and innovation. By leveraging data-driven insights and facilitating collaboration across disciplines and institutions, these practices hold the potential to transform healthcare from a reactive model to a proactive and patient-centric endeavor that improves lives, empowers practitioners, and contributes to the advancement of public health on a global scale.

6.5 IOT AND BLOCKCHAIN TECHNOLOGY IN CUTTING-EDGE METHODS FOR DATA SHARING AND PRIVACY

By utilizing cutting-edge methods, the Internet of Things (IoT) improves smart manufacturing processes that increase production. The sophisticated and intricate nature of the manufacturing process is due to the numerous IoT domains (such as industries). A finished product is the result of the work of numerous departments from various businesses. However, this raises concerns about the security and privacy of cross-domain communication. A difficult study area is cross-domain data sharing for product manufacturing [19]. A platform for cross-domain data sharing via different security gateways is proposed. The information is stored in the centralized cloud by the security gateways using the blockchain. Following a malicious activity report from the application, the centralized cloud confirms the issue via the blockchain. Additionally, the perpetrator of harmful conduct in the security gateways is subject to retaliation. The algorithms are created for data transaction and authentication. The suggested system enables secure data transfer between several domains on a worldwide scale. The outcome of the experiment shows that the proposed security and privacy framework contributes to preserving trust amongst the sectors of industry that operate together in manufacturing across domains.

In the age of personalized healthcare and the digital healthcare industry, it is crucial to gather medical data that is dispersed throughout healthcare organizations. The sharing of medical records is hampered by healthcare companies' cyberinfrastructure boundaries and potential privacy leaks. Blockchain, a decentralized public ledger known for its openness, tamper-evidence, trustlessness, and decentralization, can aid in the construction of a network for secure medical data exchange [20].

The integration of IoT technology with healthcare analytics has ushered in a new era of data-driven healthcare transformation. IoT, which involves connecting physical devices and objects to the internet to gather and exchange data, holds immense potential to revolutionize healthcare by enabling real-time monitoring, offering predictive insights, and providing personalized patient care. IoT and sophisticated healthcare analytics work together to produce a potent synergy that affects many facets of healthcare management and delivery [21]. Smart medical equipment, wearable sensors, and remote monitoring tools are examples of IoT devices that continuously create data from patients and healthcare facilities. These devices gather biometric information, medication adherence, vital signs, and other pertinent data to give a complete picture of a patient's health. Utilizing machine learning and data science methodologies, healthcare analytics scrutinizes this extensive dataset to derive significant patterns, trends, and prognostications [22].

The healthcare industry has completely changed as a result of the convergence of IoT technology and health analytics, bringing in a revolutionary period characterized by data-driven insights and individualized patient care. IoT gathers a plethora of real-time data from patients, medical devices, and healthcare environments through its networked devices and sensors [23]. Using modern health analytics, this data is subjected to in-depth analysis to identify patterns, trends, and correlations that hold the key to enhancing patient experiences, operational effectiveness, and clinical results. Healthcare providers have never-before-seen access to patient vitals and behavior through wearables with IoT capabilities and remote monitoring tools. This allows for early anomaly detection and rapid intervention. After this flood of data is processed by health analytics, useful insights are uncovered that guide treatment choices, forecast disease trajectories, and maximize resource utilization. This integration has far-reaching implications, from personalized treatment plans and remote patient management to proactive public health interventions and innovative research endeavors. By harnessing the power of IoT and health analytics, healthcare is poised to evolve into a more proactive, precise, and patient-centric domain, driving better outcomes and ultimately shaping a healthier future for individuals and communities alike [24].

6.6 PROPOSED SOLUTIONS LENSING LEGAL ASPECTS

Data quality evaluation has become a crucial step before data analytics in a data-driven economy that struggles to handle the volume and diversity of information. Data from the real world frequently has inaccuracies, conflicts, and contradictions. Such unclean data drives up processing costs and impairs analytics. When a party is considering the acquisition of data held by an unreliable institution, it is particularly crucial to evaluate the quality of a dataset. Consideration must be given to stakeholder privacy risks in this situation [25].

Regarding security and privacy protection, it is important to make sure that a patient's identity in a blockchain is kept private, especially when combined with data from other sources, which is particularly dangerous for patients with rare diseases. Additionally, the encryption must be strong enough to prevent malicious attacks, as has been reported with exchange platforms linked to cryptocurrency blockchains. With regard to limiting access, the private keys used for data encryption need specific consideration. Regulations on healthcare data privacy compliance must be maintained. With careful software development procedures and permissioned policies, security vulnerabilities can be reduced [26].

There is no specific international law exclusively dedicated to secure data sharing and collaboration in healthcare analytics. However, several international legal frameworks, conventions, and guidelines touch upon aspects of data sharing, privacy, and collaboration in the healthcare sector. These tools lay the groundwork for guaranteeing secure data practices and ethical considerations in healthcare analytics worldwide. Among them are notable international legal instruments and frameworks like the GDPR, a comprehensive data protection regulation that applies to the European Union (EU) and the European Economic Area (EEA). It mandates stringent requirements for the processing and sharing of personal data, which includes healthcare data. Organizations handling healthcare data of individuals in the EU/EEA must adhere to GDPR principles, including data minimization, purpose limitation, and obtaining explicit consent. Developed by the Council of Europe, this convention focuses on the protection of personal data, including healthcare data, in the context of automated processing. It establishes principles for data sharing and cross-border data flows while emphasizing the importance of privacy and data security [27]. The WHO ethical guidelines for health-related research provide a framework for ethical considerations in health-related research, including data sharing and collaboration. While not legally binding, they influence research practices and promote responsible data handling in healthcare analytics. While these legal frameworks and guidelines provide valuable guidance, the rapidly evolving nature of healthcare analytics and data sharing may necessitate ongoing updates and adaptations in international law. Additionally, individual countries may have their own regulations and legal requirements for healthcare data sharing and collaboration. For the most current and comprehensive information, it is advisable to consult legal experts and authorities in the relevant jurisdictions [28].

In India, the legal landscape concerning patient privacy and the handling of healthcare-related big data is a multifaceted framework guided by several laws and regulations. The Information Technology Act, 2000, plays a pivotal role in addressing electronic health records and healthcare data by establishing norms for secure data storage, transmission, and protection. The forthcoming Personal Data Protection Bill, 2019, poised to significantly impact data governance, aims to regulate the processing of personal data, including health-related information. While awaiting enactment, the bill emphasizes consent, data localization, and individual rights, shaping a comprehensive data protection framework. The Medical Council of India (Professional Conduct, Etiquette and Ethics) Regulations, 2002, underscores the ethical responsibilities of medical practitioners, stressing patient confidentiality and the necessity of informed consent before divulging patient information. Similarly, the Indian Medical Council (Professional Conduct, Etiquette, and Ethics) Regulations, 2002, reinforce these ethical standards, harmonizing the medical community's commitment to safeguarding patient privacy [29]. The Healthcare Information Security

Practices and Procedures (HISPP) Guidelines, issued by the Ministry of Health and Family Welfare, provide essential recommendations to ensure the security and confidentiality of healthcare data, encompassing electronic health records within the purview of their directives. Furthermore, the Telemedicine Practice Guidelines, 2020, comprehensively address telemedicine's evolving landscape, including protocols for maintaining patient privacy and data security in the context of remote healthcare delivery. The National Health Data Management Policy, 2020, underscores the nation's commitment to responsible health data management, advocating for data privacy, consent, and secure sharing to facilitate research and public health initiatives [30]. As India's healthcare ecosystem continues to advance, these laws and regulations coalesce to navigate the intricate intersection of patient privacy and the transformative potential of big data, ensuring that data-driven progress aligns with ethical principles and legal safeguards.

6.7 CONCLUSION AND FUTURE SCOPE

This chapter has shed light on a number of important conclusions and offered insightful information on how blockchain technology and AI can be integrated into the healthcare industry. Combining these state-of-the-art technologies could transform healthcare in a number of ways by resolving pressing issues and opening up fresh possibilities. The preservation and security of medical data are expected to improve with the integration of blockchain technology with healthcare data management. The cryptographic protections and immutable ledger of blockchain technology can greatly reduce the vulnerabilities present in conventional data storage methods. Ensuring the integrity of medical records significantly lowers the risk of unauthorized access and data breaches. This protects individuals' private health information while also fostering confidence in healthcare systems, which is essential for the delivery of high-quality healthcare.

The potential for blockchain and AI to work together to transform health analytics is exciting. Combining blockchain's data-sharing capabilities with AI's capacity to glean insightful information from large datasets allows healthcare organizations to leverage aggregated data for clinical decision support, personalized treatment, and predictive analytics. Better treatment strategies, more precise diagnoses, and ultimately better patient outcomes are made possible by this convergence. With the ability to save lives and lower healthcare costs, the potential impact on public health is immense.

There are obstacles to overcome before blockchain and AI can be successfully integrated into healthcare, particularly when it comes to ethical and regulatory issues. The privacy, ownership, and consent of medical data are critical issues that require serious consideration. Healthcare firms must prioritize ethical and transparent data management practices, and regulatory frameworks must adapt to keep up with technology changes. It is impossible to overestimate the significance of patient consent and data security, and players from all facets of the healthcare ecosystem need to collaborate to create thorough rules that respect these values.

The integration of blockchain with AI has huge potential to improve healthcare by providing more individualized, secure, and efficient services. But in order to fully achieve this promise, we must first aggressively address the ethical and legal issues raised by new technologies. Blockchain and AI technology have the potential to significantly change the healthcare industry, but they also put a great deal of responsibility on stakeholders to protect patient privacy, security, and ethical standards.

It is critical that, going forward, legislators, tech developers, healthcare practitioners, and legal experts collaborate to establish a strong and fair framework for incorporating blockchain and AI into healthcare. The utmost ethical standards, patient privacy, and well-being must all be adhered to in doing this. They will help fully realize the transformative potential of these technologies and guarantee that healthcare continues to be an area of trust, compassion, and quality.

REFERENCES

[1] Liang, X., Zhao, J., Shetty, S., Liu, J., & Li, D. (2017, October). Integrating blockchain for data sharing and collaboration in mobile healthcare applications. In *2017 IEEE 28th Annual International Symposium on Personal, Indoor, and Mobile Radio Communications (PIMRC)* (pp. 1–5). IEEE.

[2] Singh, P., Masud, M., Hossain, M. S., & Kaur, A. (2021). Cross-domain secure data sharing using blockchain for industrial IoT. *Journal of Parallel and Distributed Computing, 156*(1), 176–184.

[3] Jin, H., Luo, Y., Li, P., & Mathew, J. (2019). A review of secure and privacy-preserving medical data sharing. *IEEE Access, 7*(1), 61656–61669.

[4] Singh, B. (2023). Blockchain technology in renovating healthcare: Legal and future perspectives. In *Revolutionizing Healthcare Through Artificial Intelligence and Internet of Things Applications* (pp. 177–186). IGI Global.

[5] Hickman, C. F. L., Alshubbar, H., Chambost, J., Jacques, C., Pena, C. A., Drakeley, A., & Freour, T. (2020). Data sharing: Using blockchain and decentralized data technologies to unlock the potential of artificial intelligence: What can assisted reproduction learn from other areas of medicine?. *Fertility and Sterility, 114*(5), 927–933.

[6] Fabian, B., Ermakova, T., & Junghanns, P. (2015). Collaborative and secure sharing of healthcare data in multi-clouds. *Information Systems, 48*(1), 132–150.

[7] Wang, S., Jiang, X., Tang, H., Wang, X., Bu, D., Carey, K., ... & Ohno-Machado, L. (2017). A community effort to protect genomic data sharing, collaboration and outsourcing. *NPJ Genomic Medicine, 2*(1), 33.

[8] Welten, S., Hempel, L., Abedi, M., Mou, Y., Jaberansary, M., Neumann, L., & Kirsten, T. (2022). Multi-institutional breast cancer detection using a secure on-boarding service for distributed analytics. *Applied Sciences, 12*(9), 4336.

[9] King, R. J., Garrett, N., Kriseman, J., Crum, M., Rafalski, E. M., Sweat, D., ... & Cutts, T. (2016). Peer Reviewed: A Community Health Record: Improving Health Through Multisector Collaboration, Information Sharing, and Technology. *Preventing Chronic Disease, 13*, 10–30.

[10] Chakraborty, C. (Ed.). (2022). *Digital Health Transformation with Blockchain and Artificial Intelligence*. CRC Press.

[11] Yang, J. J., Li, J. Q., & Niu, Y. (2015). A hybrid solution for privacy preserving medical data sharing in the cloud environment. *Future Generation Computer Systems, 43*(1), 74–86.

[12] Knaup, P., Deserno, T. M., Prokosch, H. U., & Sax, U. (2018). Implementation of a national framework to promote health data sharing. *Yearbook of Medical Informatics, 27*(01), 302–304.

[13] Tandon, A., Dhir, A., Islam, A. N., & Mäntymaki, M. (2020). Blockchain in healthcare: A systematic literature review, synthesizing framework and future research agenda. *Computers in Industry, 122*, 103290.

[14] Shah, T., Wilson, L., Booth, N., Butters, O., McDonald, J., Common, K., & Murtagh, M. (2019). Information-sharing in health and social care: Lessons from a socio-technical initiative. *Public Money & Management, 39*(5), 359–363.

[15] Lu, Y., Huang, X., Zhang, K., Maharjan, S., & Zhang, Y. (2020). Blockchain empowered asynchronous federated learning for secure data sharing in internet of vehicles. *IEEE Transactions on Vehicular Technology, 69*(4), 4298–4311.

[16] Tse, D., Chow, C. K., Ly, T. P., Tong, C. Y., & Tam, K. W. (2018, August). The challenges of big data governance in healthcare. In *2018 17th IEEE International Conference On Trust, Security And Privacy In Computing And Communications/12th IEEE International Conference On Big Data Science And Engineering (TrustCom/BigDataSE)* (pp. 1632–1636). IEEE.

[17] Muheidat, F., & Tawalbeh, L. A. (2021). Artificial intelligence and blockchain for cybersecurity applications. In *Artificial Intelligence and Blockchain for Future Cybersecurity Applications* (pp. 3–29). Cham: Springer International Publishing.

[18] Kim, J. W., Edemacu, K., & Jang, B. (2019). MPPDS: multilevel privacy-preserving data sharing in a collaborative eHealth system. *IEEE Access, 7*, 109910–109923.

[19] Abbas, A., Alroobaea, R., Krichen, M., Rubaiee, S., Vimal, S., & Almansour, F. M. (2021). Blockchain-assisted secured data management framework for health information analysis based on Internet of Medical Things. *Personal and Ubiquitous Computing*, 1–14.

[20] Liang, X., Shetty, S., Tosh, D., Bowden, D., Njilla, L., & Kamhoua, C. (2018). Towards blockchain empowered trusted and accountable data sharing and collaboration in mobile healthcare applications. *EAI Endorsed Transactions on Pervasive Health and Technology, 4*(15).

[21] Gami, B., Agrawal, M., Mishra, D. K., Quasim, D., & Mehra, P. S. (2023). Artificial intelligence-based blockchain solutions for intelligent healthcare: A comprehensive review on privacy preserving techniques. *Transactions on Emerging Telecommunications Technologies, 34*(9), e4824.

[22] Allen, C., Des Jardins, T. R., Heider, A., Lyman, K. A., McWilliams, L., Rein, A. L., ... & Turske, S. A. (2014). Data governance and data sharing agreements for community-wide health information exchange: lessons from the beacon communities. *EGEMS, 2*(1).

[23] Hasan, K., Chowdhury, M. J. M., Biswas, K., Ahmed, K., Islam, M. S., & Usman, M. (2022). A blockchain-based secure data-sharing framework for Software Defined Wireless Body Area Networks. *Computer Networks, 211*, 109004.

[24] Jayasudha, M., & Vijayalakshmi, C. (2022, April). Blockchain meets healthcare: Architecture for secure data sharing in unobtrusive medical applications. In *AIP Conference Proceedings*, 2405(1), AIP Publishing.

[25] Soni, M., & Singh, D. K. (2023). A key exchange system for secure data coordination in healthcare systems. *Healthcare Analytics, 3*, 100138.

[26] Preetha, A. D., & Kumar, T. P. (2021). Securing IoT-based healthcare systems from counterfeit medicine penetration using Blockchain. *Applied Nanoscience*, 1–13.

[27] Onik, M. M. H., Aich, S., Yang, J., Kim, C. S., & Kim, H. C. (2019). Blockchain in healthcare: Challenges and solutions. In *Big data analytics for intelligent healthcare management* (pp. 197–226). Academic Press.

[28] Yogeshwar, A., & Kamalakkannan, S. (2022, March). Building dynamic permutation based privacy preservation model with block chain technology for IoT healthcare sector. In *2022 International Conference on Advanced Computing Technologies and Applications (ICACTA)* (pp. 1–8). IEEE.

[29] Ahmed, Z., Mohamed, K., Zeeshan, S., & Dong, X. (2020). Artificial intelligence with multi-functional machine learning platform development for better healthcare and precision medicine. *Database, 2020*, baaa010.

[30] Reegu, F. A., Abas, H., Gulzar, Y., Xin, Q., Alwan, A. A., Jabbari, A., & Dziyauddin, R. A. (2023). Blockchain-based framework for interoperable electronic health records for an improved healthcare system. *Sustainability, 15*(8), 6337.

CHAPTER 7

From Industry 4.0 to 5.0

Navigating the Human-Tech Symbiosis in the Era of COVID-19

Kishor Kumar Reddy C, Nishitha Pujala, and Srinath Doss

7.1 INTRODUCTION

Industry 4.0, also known as the Fourth Industrial Revolution, is a concept that brought advanced technologies and digital systems into the health sector. Industry 4.0 extends beyond health to revolutionize manufacturing, distribution, and usage processes through advanced technologies and digital systems. This transformation aimed to create "smart factories" and revolutionize the way products are made, distributed, and used. It is a departure from traditional methods, embracing interconnected devices, automation, and data-driven processes [1].

The main technologies behind Industry 4.0 are:

- Internet of Things (IoT): Devices, sensors, and machines connected through the internet, allowing them to share real-time data and communicate.
- Big Data and Analytics: Colossal amounts of data collected from connected devices analyzed to gain insights and optimize processes.
- Artificial Intelligence (AI) and Machine Learning: AI systems analyze complex data patterns, predict outcomes, and optimize processes automatically.
- Cloud Computing: Manufacturing data and processes stored in the cloud, enabling remote access and collaboration.
- Autonomous Robots: AI-powered robots perform tasks autonomously, improving efficiency and safety.

DOI: 10.1201/9781032648323-7

- Augmented Reality (AR) and Virtual Reality (VR): These technologies overlay digital information on the physical world, aiding in training and collaboration

The technologies of Industry 4.0 are shown in Figure 7.1.

Industry 4.0 remains a transformative force in manufacturing, emphasizing agility, efficiency, and customer-centricity [2]. It hinges on data-driven strategies, predictive maintenance, and tailored product offerings. At its core is the rapid evolution of the IoT, which interlinks devices and sensors to facilitate real-time data exchange. This connectivity empowers factories with profound operational insights, yielding heightened efficiency, reduced downtimes, and more informed decision-making.

Digital twins, another remarkable advancement, create virtual replicas of physical assets and processes [3]. These digital counterparts allow manufacturers to simulate different scenarios, enabling testing, optimization, and innovation without impacting the physical environment. This technology is particularly valuable for complex processes, as it reduces the risk of costly errors and supports continuous improvements. Furthermore, the integration of AR and VR enhances human-machine interaction and workforce training. AR and VR applications provide intuitive ways for employees to visualize and

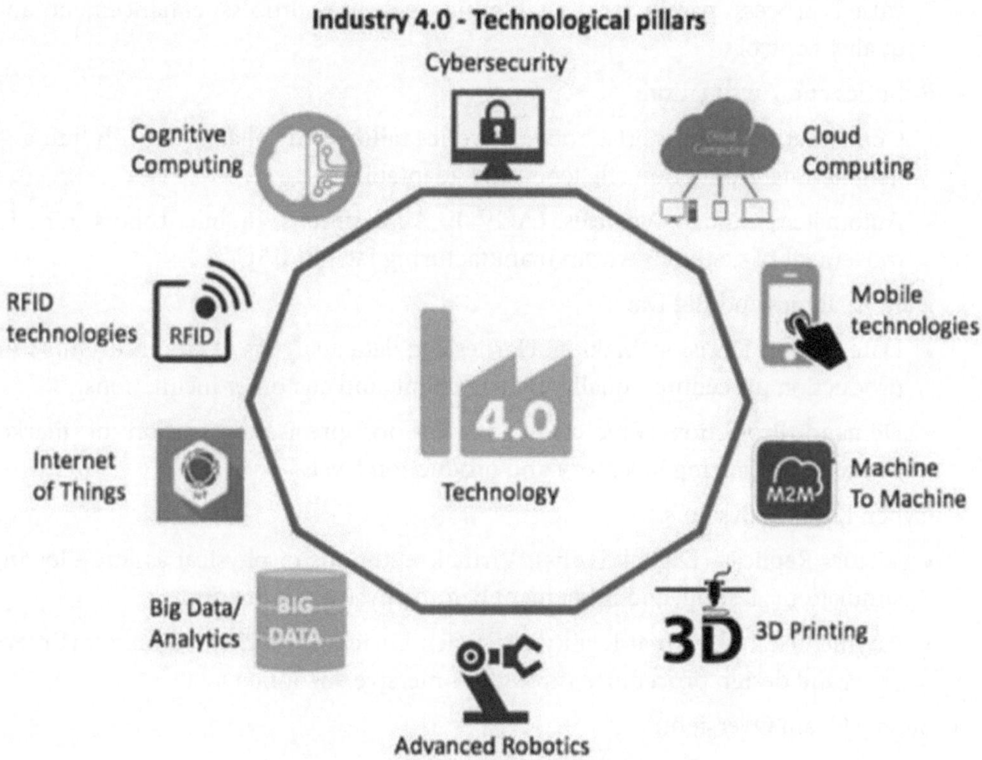

FIGURE 7.1 Technologies for Industry 4.0.

interact with complex systems, aiding in maintenance, troubleshooting, and even remote collaboration. As these technologies continue to mature, they contribute to increased productivity, reduced training times, and improved operational accuracy [4].

In conclusion, Industry 4.0 is marked by advances across various technological fronts. The fusion of IoT, AI, ML, digital twins, and AR/VR is reshaping manufacturing by enabling real-time insights, predictive capabilities, streamlined operations, and improved human–machine interactions. These are some of the applications and advantages of Industry 4.0 technologies, such as digital twins and AR/VR integration, which enable simulation, optimization, and improved human-machine interaction in manufacturing.

- Intelligent Manufacturing and Internet of Things:
  - Anticipatory Maintenance: Employing sensors and data analysis to oversee equipment conditions and forecast maintenance requirements, consequently curtailing downtime and refining maintenance schedules.
  - Resource Tracking: Real-time monitoring of resources, materials, and items within the production cycle, amplifying visibility and effectiveness in the supply chain.
  - Enhanced Sensing Technology: Integrating sensors for data aggregation on varied process parameters, facilitating real-time process enhancement and quality control.
- Robotics and Automation:
  - Collaborative Robotics (Cobots): Robotics tailored to collaborate with humans, heightening production efficiency and adaptability.
  - Automated Guided Vehicles (AGVs): Autonomous mobile robots for the movement of materials within manufacturing facilities[5].
- Data Analytics and Big Data:
  - Data-Driven Decision-Making: Harnessing data analytics to attain insights into production procedures, quality management, and customer inclinations.
  - Demand Projection: Predictive analytics for precise estimation of market demand, optimizing inventory and production levels.
- Cyber-Physical Systems:
  - Virtual Replicas (Digital Twins): Virtual renditions of physical assets, allowing simulations, testing, and refinement before physical production.
  - Augmented and Virtual Reality (AR/VR): Enriching worker training, maintenance, and design procedures through immersive encounters.
- Supply Chain Oversight:
  - Smart Logistics: Streamlining goods movement via interconnected systems, thereby reducing transportation expenses and augmenting efficiency.
  - Supplier Synergy: Real-time data exchange and synchronization with suppliers, guaranteeing punctual deliveries and curtailing disruptions.

These innovations are revolutionizing manufacturing by fostering real-time data insights, autonomous decision-making, risk-free experimentation, and immersive human-machine interaction. As businesses increasingly adopt these technologies, they are poised to achieve higher levels of efficiency, flexibility, and innovation in their operations. Industry 4.0 involves a diverse range of stakeholders who collectively contribute to its evolution and adoption. At the heart of this transformation are manufacturing companies and industries themselves. They are the primary beneficiaries of Industry 4.0 technologies, using IoT, AI, and automation to optimize production, reduce costs, enhance product quality, and create more customized offerings [7].

The development within Industry 4.0 continues to evolve at a rapid pace, shaping the way industries operate and transforming traditional manufacturing approaches. One significant trend is the increasing integration of artificial intelligence (AI) and machine learning (ML) into manufacturing processes. Digital twins are evolving to become more sophisticated, offering highly accurate virtual replicas of physical assets and systems. Furthermore, the convergence of these technologies is driving the development of smart factories and ecosystems Collaborative robots (cobots) are becoming more advanced, working alongside human operators to enhance productivity and safety [8]. AR and VR are being used for training, maintenance, and remote assistance, bridging the gap between the physical and digital realms.

7.2 INDUSTRY 4.0 TO INDUSTRY 5.0

Industry 4.0 represents a transformative integration of technology into society, where new advancements connect the physical, digital, and biological worlds, impacting all disciplines, economies, and industries. The shift from Industry 4.0 to Industry 5.0 marks a significant transformation in the manufacturing and industrial landscape. The key distinctions and characteristics of Industry 5.0 compared to its predecessor are as follows:

- Human-Tech Collaboration: Industry 5.0 acknowledges that the true potential of manufacturing lies in the seamless collaboration between humans and machines. Instead of replacing humans with automation, Industry 5.0 aims to empower and augment human capabilities using technology [11].
- Personalization and Customization: Industry 5.0 facilitates mass customization and personalized production. Through the application of AI and digital twins, manufacturers can customize products to meet individual customer preferences.
- Decentralization and Flexibility: Industry 5.0 moves away from centralized and inflexible production systems. Decentralized manufacturing networks enable greater flexibility and responsiveness to changing market demands
- Sustainability Focus: Industry 5.0 places a strong emphasis on sustainability and eco-friendly practices. The integration of green technologies and circular economy principles help minimize environmental impacts.

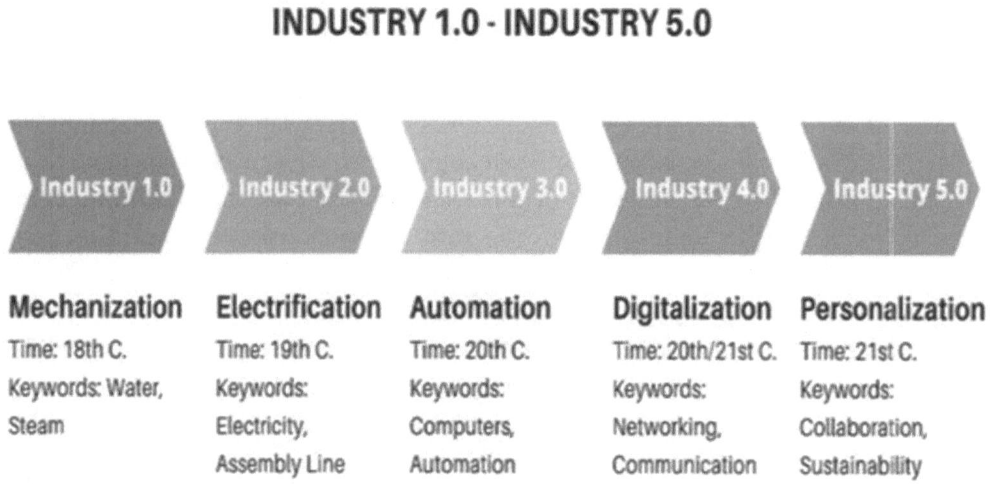

FIGURE 7.2 A Look at Industry 1.0 to Industry 5.0.

- Advanced Human-Machine Interfaces: Industry 5.0 incorporates intuitive interfaces that enable seamless communication between humans and machines, fostering enhanced cooperation.

Industry 5.0 represents a new era in manufacturing, where technology complements human labor rather than replacing it [12]. The integration of human creativity and emotional intelligence with cutting-edge technology paves the way for a more productive, sustainable, and inclusive industrial future. The transition from Industry 1.0 to Industry 5.0 is shown in Figure 7.2.

7.3 WHY INDUSTRY 5.0?

Industry 5.0 marks a revolutionary advancement in the manufacturing and industrial sector, building on the progress achieved during Industry 4.0. This transformation is driven by a combination of factors aimed at overcoming the limitations and challenges encountered in previous industrial revolutions, while capitalizing on the opportunities presented by emerging technologies. Primarily, Industry 5.0 underscores a human-centric approach, recognizing the importance of technology in empowering and enhancing human capabilities rather than replacing human workers. By fostering collaboration between humans and machines, this new model encourages the integration of human creativity, problem-solving skills, and emotional intelligence with advanced technologies, leading to a more productive and innovative manufacturing environment [13].

Additionally, Industry 5.0 places significant emphasis on personalization and customization. Employing technologies like AI, digital twins, and smart manufacturing, businesses can cater to individual customer preferences effectively. This ability for mass customization allows manufacturers to create products tailored to unique customer needs, thereby enhancing customer satisfaction and loyalty. Moreover, Industry 5.0 promotes flexibility and agility in manufacturing processes by adopting decentralized manufacturing networks. This enables companies to respond swiftly to changing market demands and dynamic customer preferences, fostering improved responsiveness and adaptability that keeps businesses competitive in fast-paced industries.

Furthermore, Industry 5.0 is deeply committed to sustainability and green practices. Addressing concerns about environmental impacts and resource depletion, this new era incorporates green technologies and circular economy principles. The objective is to minimize waste, energy consumption, and environmental footprints, contributing to a more sustainable future [14]. The transition to Industry 5.0 necessitates a skilled and versatile workforce. As automation and intelligent machines become more integral to manufacturing processes, the importance of continuous training and up skilling programs becomes evident. Empowering the workforce with the requisite knowledge and skills ensures that human talent can effectively collaborate with advanced technologies, leading to greater efficiency and creativity.

In conclusion, Industry 5.0 signifies a profound shift in manufacturing, embracing a human-centric, personalized, and sustainable approach to industrial development. By harnessing advanced technologies while fostering collaboration between humans and machines, this new era aims to create a more innovative, flexible, and ethical industrial future. Empowering the workforce with the necessary skills and knowledge and promoting global collaboration are critical components for the success of Industry 5.0 [15].

7.4 APPLICATIONS IN INDUSTRY 5.0

Industry 5.0, characterized by its focus on human-machine collaboration and advanced technologies, presents a wide array of applications that hold transformative potential across diverse sectors. The following are some key examples of Industry 5.0 applications:

- Intelligent Manufacturing: Industry 5.0 facilitates smart manufacturing processes where humans and machines collaborate to optimize production. Collaborative robots, also known as cobots, can aid workers in physically demanding or repetitive tasks, leading to increased productivity and improved worker safety.
- Customized Production: Through the integration of AI and digital twins, Industry 5.0 enables mass customization and personalized production. Manufacturers can create products tailored to individual customer preferences, offering enhanced customer experiences [16].
- Decentralized Production Networks: Industry 5.0 encourages the adoption of decentralized production networks, enabling interconnected factories to communicate and collaborate in real-time. This approach enhances flexibility and agility in responding to dynamic market demands.
- Healthcare Advancements: In the healthcare sector, Industry 5.0 technologies can support collaborative robots and AI-driven diagnostics, assisting medical

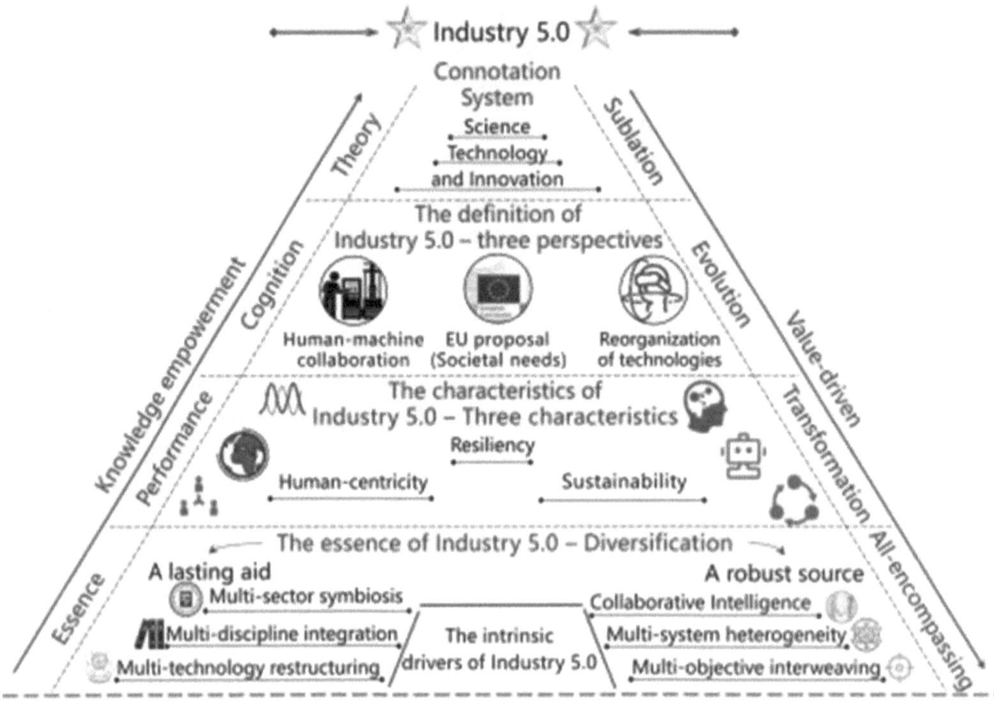

FIGURE 7.3 Industry 5.0: Prospect and Retrospect.

professionals in surgeries, patient care, and diagnosis, ultimately leading to improved healthcare outcomes.

- Enhanced Supply Chain Management: Industry 5.0 extends its applications to supply chain management, where interconnected systems enable real-time tracking of goods, predictive maintenance of machinery, and optimized logistics for enhanced efficiency and cost reduction.
- Precision Agriculture: Industry 5.0 finds application in agriculture, particularly in precision farming. Utilizing sensors, drones, and AI technologies, farmers can monitor crops, optimize irrigation, and apply fertilizers with precision, leading to increased yields and reduced environmental impact.

To conclude, Industry 5.0 presents a wide spectrum of applications across various sectors, fostering innovation, efficiency, and sustainability. The synergistic integration of human creativity with advanced technologies unlocks transformative possibilities that can reshape industries and elevate the overall quality of life. The applications of Industry 5.0 are shown in Figure 7.3.

7.5 TECHNOLOGIES IN INDUSTRY 5.0

Industry 5.0 integrates cutting-edge technologies that build on Industry 4.0's foundation while emphasizing collaboration between humans and machines, ethical considerations, and societal impact. These technologies work in harmony to create a more unified and adaptable ecosystem [17].

- Advanced Artificial Intelligence (AI) and Machine Learning (ML): In Industry 5.0, AI and ML have greater complexity. These technologies empower machines not only to analyze data but also to comprehend context and intent, facilitating more natural interactions with humans.
- Extended Reality (XR): Extended reality, encompassing virtual reality (VR), augmented reality (AR), and mixed reality (MR), is a cornerstone of Industry 5.0. These technologies bridge the gap between the physical and digital domains, facilitating immersive experiences and collaborative settings.
- Blockchain and Secure Data Management: Blockchain technology offers transparent, immutable records of transactions and data transfers, cultivating trust and traceability in supply chains, financial dealings, and more.
- Quantum Computing: Industry 5.0 introduces quantum computing as a transformative factor. Quantum computers have the potential to process complex calculations exponentially faster than conventional computers, unlocking opportunities to solve intricate optimization problems, simulate complex phenomena, and refine machine learning algorithms [18].

7.6 CHALLENGES IN INDUSTRY 5.0

Industry 5.0, which marks the next stage in the evolution of industrial manufacturing, introduces a significant shift that centers on collaboration between humans and machines. While it presents promising advancements and opportunities, this transition is accompanied by its own set of hurdles that must be overcome to ensure successful implementation. One of the primary obstacles is the adaptability of the workforce [19]. As Industry 5.0 integrates advanced automation and intelligent machines, conventional job roles may undergo transformation or even become obsolete. Consequently, the workforce needs to adjust to these changes, necessitating upskilling and reskilling initiatives.

To achieve this, companies must invest in comprehensive training programs to equip employees with the essential knowledge and competencies to effectively work alongside machines and leverage new technologies. Ethical considerations play a crucial role in the implementation of Industry 5.0. With the increasing use of AI and robotics, concerns related to data privacy, algorithmic bias, and accountability become more pronounced. Striking a balance between efficiency and ethical practices becomes imperative.

Data security and privacy become paramount in the context of Industry 5.0. The interconnected nature of devices and systems exposes an extensive attack surface for potential cyber threats. To safeguard sensitive data and protect against cyber-attacks, companies must adopt stringent cybersecurity measures. The potential consequences of data breaches could have far-reaching impacts on both businesses and consumers. Interoperability and standardization represent critical challenges in Industry 5.0.

The cost of implementation can also pose a barrier to the widespread adoption of Industry 5.0, particularly for small- and medium-sized enterprises. The initial investment in new equipment, infrastructure, and workforce training may be substantial. As a result, businesses and governments need to explore cost-effective approaches, such as public-private partnerships and incentives, to facilitate smoother transitions. Overcoming resistance to change is an organizational challenge.

Lastly, adapting regulatory frameworks to accommodate the unique aspects of Industry 5.0 is crucial. Traditional regulations may not adequately address the complexities and ethical implications of human-machine interactions. Regulatory bodies must collaborate with industry experts to develop new guidelines that ensure the safe, ethical, and responsible deployment of Industry 5.0 technologies. In conclusion, Industry 5.0 holds immense promise for the manufacturing sector, but it also presents a set of challenges that require proactive measures to overcome.

7.7 CASE STUDY

Industry 5.0: A Remedy for the Covid-19 Pandemic Phase

7.7.1 Objectives

- To examine the impact of Industry 5.0 in the Health Care, Education, and Banking sectors amid the COVID-19 pandemic.
- To analyze the challenges involved in integrating Industry 5.0.
- To explore the diverse supportive features and capabilities of Industry 5.0 in addressing the COVID-19 pandemic.

7.7.2 Industry 4.0 Response

At the outset of the COVID-19 pandemic, Industry 4.0 technologies played a pivotal role in enabling rapid response strategies. Governments and health organizations used AI-powered models to forecast potential infection rates and plan resource allocation. Machine learning algorithms processed vast amounts of data to identify potential treatments, contributing to the expedited identification of drug candidates.

Amidst the pandemic, Industry 4.0 technologies facilitated remote patient monitoring using IoT devices as follows:

- IoT-Enabled Remote Monitoring: As the pandemic unfolded, Industry 4.0 technologies extended into healthcare delivery. Remote patient monitoring through IoT devices gained prominence, allowing healthcare professionals to track patients' health conditions remotely.
- Robotics in Sanitization and Delivery: In Industry 4.0, robots were employed for sanitization and logistics. Drones and robots were used to disinfect public spaces and deliver medical supplies to hospitals. These autonomous systems helped reduce human exposure to the virus while maintaining critical supply chains.

- Telemedicine and Virtual Care: Telemedicine saw a surge in adoption during Industry 4.0. Patients sought medical consultations remotely through video calls and chat platforms [19]. This approach eased the strain on healthcare facilities and provided patients with access to medical expertise without in-person physical visits.
- Transition to Industry 5.0: The transition to Industry 5.0 was characterized by a convergence of technologies and a deeper emphasis on human-machine collaboration, ethics, and patient-centric care.
- AI-Driven Drug Discovery: In Industry 5.0, AI-powered drug discovery accelerated the search for treatments and vaccines. Machine learning algorithms analyzed vast datasets to identify potential drug candidates and predict their effectiveness. This approach drastically shortened the time required for drug development, demonstrating the potential for AI to transform healthcare research.
- Human-Robot Collaboration: Collaborative robotics or cobots became a key asset in Industry 5.0's response. Cobots worked alongside healthcare professionals, assisting in tasks like sample collection, patient monitoring, and disinfection. This human-robot collaboration enhanced safety, optimized resource allocation, and showed the potential for harmonious man-machine partnerships.
- Blockchain-Enabled Vaccine Management: As COVID-19 vaccines became available, Industry 5.0 integrated blockchain technology for secure and transparent vaccine distribution. Blockchain ensured the integrity of supply chains, tracked vaccine shipments and storage conditions, and provided a tamper-proof record of vaccination.
- Ethical AI in Contact Tracing: Ethical considerations were paramount in Industry 5.0's approach to contact tracing. AI-driven contact tracing systems prioritized privacy through anonymization techniques and secure data handling. By striking a balance between public health needs and individual privacy rights, ethical AI contributed to effective pandemic management [20]. The transition from Industry 4.0 to Industry 5.0 due to impact of COVID-19 is shown in Table 7.1.

TABLE 7.1 Transition from Industry 4.0 to Industry 5.0: COVID-19 Impact

Aspect	Industry 4.0 (Pre-COVID)	Industry 5.0 (During/Post-COVID)
Investments	$X billion	$Y billion
Automation Level	High	Moderate-High
Remote Work	Limited	Widely Adopted
Supply Chain	Global	Localized/Resilient
Human Involvement	Moderate	Significant
Focus	Efficiency	Innovation/Collaboration
Communication	M2M	H2H, H2M
Skills Needed	Technical	Creative, Collaborative
Ecosystems	Data Monetization	Co-Creation, Sustainability
Sustainability	Resource Optimization	Ethical Considerations

Table 7.1 outlines the transition from Industry 4.0 to Industry 5.0 with a focus on the impact of the COVID-19 pandemic. Here's a brief description of each aspect:

Investments: In Industry 4.0 (pre-COVID), investment was around $X billion, whereas in Industry 5.0 (during/post-COVID), it shifted to approximately $Y billion.

Automation Level: Industry 4.0 was characterized by high levels of automation, while Industry 5.0 showed a transition to a moderate-to-high level of automation.

Remote Work: Pre-COVID, remote work was limited in Industry 4.0, but during and after the pandemic (Industry 5.0), remote work became widely adopted.

Supply Chain: Industry 4.0 had a global supply chain approach, while Industry 5.0 shifted towards localized and resilient supply chains.

Human Involvement: Industry 4.0 had a moderate level of human involvement, but in Industry 5.0, human involvement became significantly more prominent.

Focus: In Industry 4.0, the primary focus was on efficiency improvements, whereas Industry 5.0 emphasized innovation and collaboration.

Communication: Industry 4.0 relied heavily on machine-to-machine (M2M) communication, while Industry 5.0 emphasized more human-to-human (H2H) and human-to-machine (H2M) communication.

Skills Needed: Industry 4.0 demanded mainly technical skills, whereas Industry 5.0 required a shift towards creative and collaborative skills.

Ecosystems: Industry 4.0 ecosystems were centered around data monetization, whereas Industry 5.0 ecosystems shifted towards co-creation and sustainability.

Sustainability: While Industry 4.0 focused on resource optimization, Industry 5.0 placed greater emphasis on ethical considerations in addition to sustainability.

The transition from Industry 4.0 to Industry 5.0, accelerated by the COVID-19 pandemic, has led to changes in investment trends, automation, remote work, supply chain strategies, and sustainability priorities. Industry 5.0, emerging as a response to the pandemic, employs advanced technologies and collaboration to address its challenges and transform sectors in the evolving landscape. In healthcare, Healthcare 5.0 is a notable innovation, focusing on tailored solutions over standardized ones. This approach utilizes smart technology for personalized treatment, especially relevant for COVID-19 patients. Additionally, Industry 5.0 advancements assist in healthcare training for patients, doctors, and medical students.

Major technology companies are actively engaging in research and development efforts, particularly in the realm of digital wellness, with innovations such as Apple smart watches driving healthcare advancements. In essence, Healthcare 5.0 envisions a society where everyone can create value anytime, anywhere, with a focus on safety and harmony with nature. In this society, patients can transcend the limitations that previous healthcare approaches, up until Healthcare 4.0, could not overcome.

7.8 CONCLUSION

In summary, this chapter explored the transformation of the manufacturing sector from a focus on technology-centric processes to a more comprehensive model that highlights the synergy between human capabilities and advanced technologies. This transition, characterized by the concept of human-tech symbiosis, aims to merge human creativity, decision-making, and adaptability with the precision and efficiency of technologies like AI and IoT. The chapter underscored the advantages of this shift, such as heightened customization, agility, and innovation. By fostering a harmonious coexistence of human expertise and technological advancement processes can better cater to dynamic market demands and individual customer preferences.

Transitioning to Industry 5.0 was exemplified during the COVID-19 response, showcasing the integration of AI-guided drug discovery, collaborative robotics, blockchain-based vaccine management, ethically driven AI in contact tracing, and patient-centric virtual healthcare. These technologies converged to establish a more resilient, adaptable, and human-centered crisis management approach. This evolution vividly demonstrated the potential of Industry 5.0 in revolutionizing healthcare delivery and enhancing global pandemic response strategies, serving as a testament to the transformative possibilities that arise from the collaboration between human capabilities and cutting-edge technologies in critical scenarios.

BIBLIOGRAPHY

[1] Alojaiman, Bader. (2023). Technological Modernizations in the Industry 5.0 Era: A Descriptive Analysis and Future Research Directions. *Processes*.

[2] Leelavathi, R., Philip, Bijin, Babu N., Aiswarya & Mukthar Kp, Jaheer. (2023). Industry 5.0: A Panacea in the Phase of Covid-19 Pandemic Concerning Health, Education, and Banking Sector. In *International Conference on Business and Technology* (pp. 3–12). Cham: Springer International Publishing.

[3] Crnjac Žižić, Marina, Mladineo, Marko, Gjeldum, Nikola, & Celent, Luka. (2022). From Industry 4.0 towards Industry 5.0: A Review and Analysis of Paradigm Shift for the People, Organization and Technology. *Energies*.

[4] Pizoń, Jakub & Gola, Arkadiusz. (2023). Human-Machine Relationship-Perspective and Future Roadmap for Industry 5.0 Solutions.

[5] Alves, Joel, Lima, Tânia & Gaspar, Pedro. (2023). Is Industry 5.0 a Human-Centred Approach? A Systematic Review. *Processes*.

[6] Adel, Amr. (2022). Future of industry 5.0 in society: human-centric solutions, challenges and prospective research areas. *Journal of Cloud Computing*.

[7] Jafari, Niloofar, Azarian, Mathew & Yu, Hao. (2022). Moving from Industry 4.0 to Industry 5.0: What Are the Implications for Smart Logistics?. *Logistics*.

[8] di nardo, Mario & Yu, Haoxuan. (2021). Special Issue "Industry 5.0: The Prelude to the Sixth Industrial Revolution". *Applied System Innovation*.

[9] Gregório Pereira, Andreia, Lima, Tânia & Charrua Santos, Fernando. (2020). Industry 4.0 and Society 5.0: Opportunities and Threats. *International Journal of Recent Technology and Engineering (IJRTE)*.

[10] Demir, Kadir & Cicibaş, Halil. (2017). Industry 5.0 and a critique of industry 4.0.

[11] Chander, Bhanu, Pal, Souvik, De, Debashis & Buyya, Rajkumar. (2022). Artificial Intelligence-based Internet of Things for Industry.

[12] Özdemir, Vural & Hekim, Nezih. (2018). Birth of Industry 5.0: Making Sense of Big Data with Artificial Intelligence, "The Internet of Things" and Next-Generation Technology Policy. *OMICS: A Journal of Integrative Biology.*

[13] Ávila-Gutiérrez, María Jesús & Suárez Fernández-Miranda, Susana & González, Francisco. (2022). Occupational Safety and Health 5.0—A Model for Multilevel Strategic Deployment Aligned with the Sustainable Development Goals of Agenda 2030. *Sustainability.*

[14] Bochenek, Bogdan & Ustrnul, Zbigniew. (2022). Machine Learning in Weather Prediction and Climate Analyses—Applications and Perspectives. Atmosphere.

[15] Anisha, P.R., Srinivasulu, Reddy K., & Surender, Reddy S. (2012). Third Party Data Protection Applied To Cloud and Xacml Implementation in the Hadoop Environment With Sparql. *IOSR Journal of Computer Engineering (IOSRJCE)* ISSN, pp.2278–0661.

[16] Narasimha Prasad, L.V., Shankar Murthy, P., & Kishor Kumar Reddy, C. (2013). Analysis of magnitude for earthquake detection using primary waves and secondary waves, *International Conference on Human Computer Interactions (ICHCI)* (pp. 1–16), IEEE.

[17] Anisha, P.R., Kishor Kumar Reddy, C., Apoorva, K., & C Meghana Mangipudi, C. (2021/4/1). Early Diagnosis of Breast Cancer Prediction using Random Forest Classifier, *IOP Conference Series: Materials Science and Engineering* (Vol. 1116, No. 1, p. 012187). IOP Publishing..

[18] Subbarayudu, B, Lalitha Gayatri, L, Sai Nidhi, P, Ramesh, P, Gangadhar Reddy, R, & Kishor Kumar Reddy, C. (2017/8). Comparative analysis on sorting and searching algorithms, *International Journal of Civil Engineering and Technology (IJCIET),* 8(8), pp. 955–78.

[19] Vijaya Babu, B, Kishor Kumar Reddy, C, & Rupa, C. H. (2015). SLGAS: Supervised learning using gain ratio as attribute selection measure to nowcast snow/no-snow. *International Review on Computers and Software.*

[20] Anisha, P.R., Reddy, C.K.K. & Prasad, L.N., (2015, January). A pragmatic approach for detecting liver cancer using image processing and data mining techniques. In *2015 International Conference on Signal Processing and Communication Engineering Systems* (pp. 352–357). IEEE.

CHAPTER 8

AI-Enabled Risk Assessment and Safety Management in Construction

Muhammad Usama, Ubaid Ullah, Zaid Muhammad, Taminul Islam, and Syeda saba Hashmi

8.1 INTRODUCTION

The construction industry plays a vital role in shaping modern infrastructure and urban landscapes but also faces significant safety and risk management challenges [1]. Construction sites are inherently complex and hazardous environments where accidents and incidents can have severe consequences for workers, the environment, and project timelines [2]. Despite efforts to improve safety standards, traditional risk assessment methods often fall short in effectively predicting and preventing accidents [3]. Integrating Artificial Intelligence (AI) in construction has shown immense promise in transforming various aspects of the industry, including safety and risk management [4]. AI technologies, such as machine learning and deep learning algorithms, can process vast amounts of data and identify patterns that human analysis may overlook [5]. By leveraging AI, construction stakeholders can make informed decisions, enhance safety measures, and minimize potential risks [6]. This chapter explores the application of AI in risk assessment and safety management within the construction sector. We will delve into the benefits and challenges of implementing AI-driven solutions and highlight the potential to revolutionize safety practices in the industry. From predictive risk analysis to real-time monitoring and incident reporting, AI offers a comprehensive approach to bolstering safety protocols and fostering a culture of proactive risk prevention. Figure 8.1 highlights the transformative impact of AI on the construction industry.

8.2 LITERATURE

Ifeanyi Okpala et al. explore the application of cutting-edge technologies in construction to enhance safety risk management. They investigate the integration of AI, machine

DOI: 10.1201/9781032648323-8

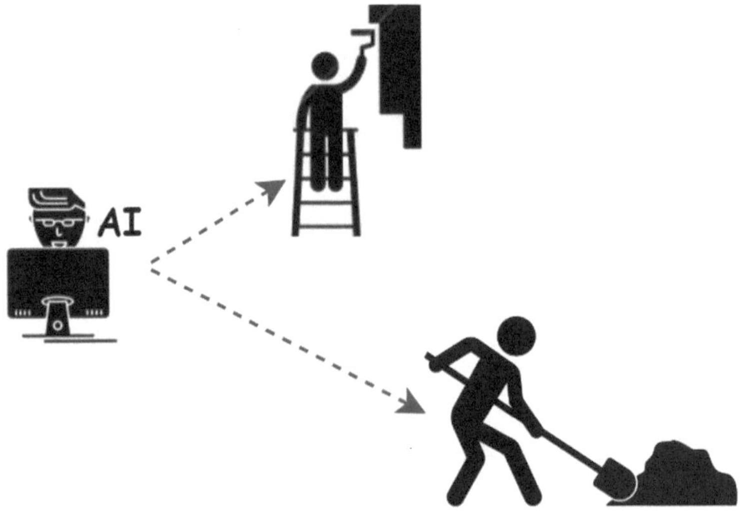

FIGURE 8.1 AI is Revolutionizing the Construction Industry.

learning, IoT, blockchain, and robotics to identify and mitigate potential hazards. By leveraging these technologies, the research aims to proactively address safety concerns, improve risk assessment accuracy, and foster a safer environment on construction sites. The study emphasizes the potential of these innovations while acknowledging challenges such as data quality, technology adoption, and model reliability [7]. Sofiat O. Abioye et al. provides an overview of the current state of AI applications in construction. They discuss potential benefits, such as improved project management and safety, along with challenges like data integration and workforce adaptation. The chapter emphasizes the need for further research and collaboration to fully harness AI's potential in construction [8].

Yue Pan et al. provide a comprehensive analysis of the application of AI in construction engineering and management. The chapter critically examines AI's various roles in improving construction processes, including risk assessment, decision support, safety monitoring, and equipment management. It reviews existing studies and highlights the potential of AI to enhance efficiency, accuracy, and safety in construction projects. Additionally, future trends and potential advancements in AI technology within the construction industry are discussed [9]. Table 8.1 provides a concise overview of the standard techniques employed in AI-Enabled Risk Assessment and Safety Management in Construction, along with descriptions and limitations.

8.3 OBJECTIVES

This chapter aims to comprehensively explore the integration of AI techniques in enhancing risk assessment and safety management practices within the construction industry. It also aims to elucidate the potential of AI-powered tools and technologies in mitigating construction-related risks, improving workplace safety, and optimizing overall project outcomes. By examining the latest advancements and case studies, the chapter aims to

TABLE 8.1 Summarizing AI-Enabled Risk Assessment and Safety Management Techniques in Construction: Insights, Applications, and Constraints

References	Technique Used	Description	Limitations
[10] [11][12]	Machine Learning algorithms, Natural Language Processing	This study applies machine learning and NLP techniques to analyze safety incident reports and predict potential risks in construction projects.	The limited dataset may lead to biased predictions—and difficulty handling unstructured data.
[13]	Artificial Intelligence, Decision Support System	The research proposes an AI-based system that assesses construction risks and provides decision support to project managers.	Relies on accurate input data may not account for unforeseen risks.
[14]	Deep Learning, Computer Vision	This study uses deep learning and computer vision to monitor construction sites for potential safety hazards and violations.	Reliability may vary based on weather conditions and image quality.
[15][16]	Predictive Analytics, Data Mining	The authors propose a predictive analytics framework to identify and mitigate safety risks in construction by analyzing historical incident data.	Requires comprehensive historical data for accurate predictions.
[17]	Internet of Things (IoT), Real-time Monitoring	This study presents an IoT-based system that monitors construction workers' safety conditions and alerts them for immediate interventions.	Relies on effective IoT infrastructure and potential privacy concerns.
[18]	Fuzzy Logic, Risk Assessment	The research utilizes fuzzy logic to assess risks in construction projects considering various qualitative and quantitative factors.	Interpretability of fuzzy logic models may be challenging. Requires expert knowledge to set up fuzzy rules.
[19]	Robotics, Simulation	This study introduces a robotic simulation approach to identify potential safety hazards and plan avoidance strategies in construction scenarios.	Simulation accuracy highly depends on the quality of input data and model assumptions.
[20]	Blockchain Technology, Safety Management	The authors propose a blockchain-based system to enhance transparency and accountability in construction safety management processes.	Requires widespread adoption of blockchain technology in the construction industry.
[11],[21]	Sensor Technology, Machine Learning	This study combines sensor data and machine learning to improve safety measures for heavy equipment operation on construction sites.	Sensor accuracy and maintenance can impact the reliability of predictions.
[22]	Human-Centric AI, Activity Recognition	The research presents an AI system that focuses on recognizing human activities to proactively address safety concerns on construction sites.	Accuracy may vary based on the complexity of human actions and environmental conditions.

thoroughly explain how AI can revolutionize traditional risk assessment methodologies and safety protocols, fostering a safer and more efficient construction environment and to present future perspectives on the evolving role of AI in enhancing safety and shaping the construction industry's safety culture. AI has the potential to be a transformative force in construction safety, empowering stakeholders to proactively address risks and create safer environments for workers and the community. As AI technologies advance, their integration into the construction industry will pave the way for safer, more efficient, and sustainable projects.

8.4 CONSTRUCTION SAFETY AND RISK MANAGEMENT

8.4.1 Importance of Safety in Construction

Safety in construction is of utmost importance as it directly impacts the well-being and lives of workers involved in the industry [23], [24]. Construction sites are inherently hazardous environments with potential risks such as falls, collapses, electrical accidents, and exposure to harmful materials [25], [26]. Prioritizing safety is essential to preventing accidents and injuries, reducing the human cost of construction projects [27]. Moreover, a safe working environment fosters a positive safety culture, encouraging workers to take personal responsibility for their well-being and that of their colleagues [28]. By implementing safety protocols and adhering to legal and regulatory requirements, construction companies can protect their workforce and improve project efficiency. Reduced accidents and injuries lead to enhanced productivity, fewer delays, and decreased downtime. This, in turn, helps maintain a positive reputation in the industry, gain client trust, and attract new opportunities. Additionally, investing in safety measures can result in cost savings by avoiding medical expenses, legal liabilities, and compensation claims. Ultimately, safety in construction is not just a moral obligation but also a strategic decision that contributes to the success of projects and the well-being of all involved parties [29], [30]. Figure 8.2 shows some of the safety signs used in the construction industry.

8.4.2 Traditional Risk Assessment Methods

Traditional risk assessment methods are systematic approaches used to identify, analyze, and evaluate potential risks within various contexts, such as business operations, projects, financial investments, etc. These methods have been developed over time and often involve structured processes to assess and mitigate risks. Some standard traditional risk assessment methods are shown in Figure 8.3.

8.4.2.1 Qualitative Risk Assessment

This method involves identifying and evaluating risks based on subjective judgments and qualitative information. Risks are typically categorized into high, medium, or low levels, and their potential impacts and likelihoods are assessed using expert opinions or historical data [31],[32].

FIGURE 8.2 Safety Signs used in Construction Works.

8.4.2.2 Quantitative Risk Assessment

Unlike qualitative assessment, this method involves assigning numerical values to various risk parameters, such as impact and likelihood. Mathematical models, statistical analysis, and historical data are used to estimate potential losses and probabilities. This approach provides more precise and quantifiable risk metrics [31], [33].

8.4.2.3 SWOT Analysis

SWOT stands for Strengths, Weaknesses, Opportunities, and Threats. It is a framework used to assess risks and opportunities within a business or project by analyzing internal strengths and weaknesses along with external opportunities and threats [34].

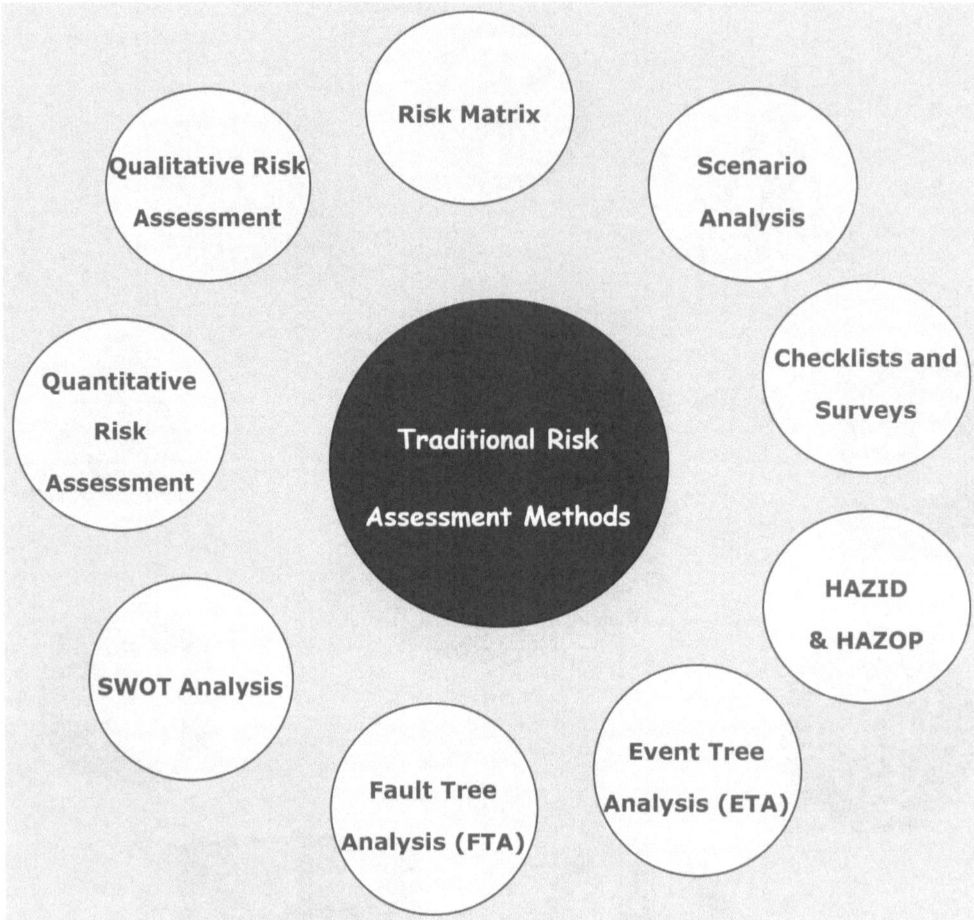

FIGURE 8.3 Traditional Risk Assessment Methods.

8.4.2.4 Fault Tree Analysis (FTA)

FTA is a deductive method used to analyze the causes of a specific event, such as an accident or system failure. It breaks down the event into contributing factors and identifies the combinations of events that can lead to the undesired outcome [35].

8.4.2.5 Event Tree Analysis (ETA)

ETA is a method that complements FTA. It focuses on the potential consequences of a specific initiating event. The analysis follows various possible outcomes and assesses their probabilities and impacts [36].

8.4.2.6 Hazard Identification (HAZID) and Hazard and Operability Study (HAZOP)

These methods are commonly used in industries such as oil and gas, chemical, and manufacturing. They involve structured brainstorming sessions to identify potential hazards, deviations from design intent, and operational risks [37].

8.4.2.7 *Checklists and Surveys*

Using predefined checklists or surveys, organizations can systematically identify and evaluate risks based on a list of predefined criteria. This method can help ensure comprehensive coverage of potential risks [38].

8.4.2.8 *Scenario Analysis*

This method involves creating different scenarios that outline various potential futures. Each scenario considers different combinations of external factors and assesses their potential impacts on the organization or project [39].

8.4.2.9 *Risk Matrix*

A risk matrix is a visual tool that helps prioritize risks based on their likelihood and potential impact. It categorizes risks into different levels of severity and guides decision-making regarding risk management strategies [40].

While traditional risk assessment methods have been widely used, they often rely on historical data and expert opinions, which may limit their ability to anticipate emerging or unprecedented risks. Modern approaches, such as probabilistic risk assessment and advanced data analytics, aim to address some of these limitations by incorporating more sophisticated modeling techniques and leveraging large datasets.

8.4.3 Limitations of Conventional Approaches

Conventional or traditional approaches to risk assessment, while widely used and proven effective in many contexts, also have some limitations that can impact their accuracy and applicability [41]. One limitation is the reliance on historical data and past experiences to assess risks. While historical data can provide valuable insights, it may not account for emerging or unprecedented risks, leaving organizations vulnerable to new and unexpected challenges [42], [43]. Another limitation is the subjectivity inherent in some conventional methods, such as risk matrix or SWOT analysis [44]. The assignment of probabilities and impacts is often based on expert judgment, which can introduce bias and inconsistency into the assessment process.

Additionally, these methods may not fully capture the complexities and interdependencies of modern systems, especially in large-scale and interconnected projects [45], [46]. Traditional approaches often focus on known risks and may overlook "unknown unknowns" or rare events with catastrophic consequences. As a result, they may not adequately address black swan events or systemic risks, which can have significant implications for organizations and society.

Moreover, conventional methods treat risks as discrete and independent entities, neglecting potential cascading effects and interactions between risks [47]. Risks are often interconnected and can amplify or mitigate each other's impacts [48]. Failing to account for these dependencies can lead to inaccurate risk assessments and inadequate risk response planning. Another limitation is the static nature of many traditional approaches. They often provide point-in-time assessments, but risks evolve, and their probabilities and impacts can change. Failure to consider dynamic risk factors may result in outdated risk profiles and ineffective risk mitigation strategies.

Furthermore, traditional methods may not adequately address the complexities of uncertainties and probabilistic risk modeling. Delays are prevalent in dynamic and complex systems, and conventional methods may not effectively capture the full range of potential outcomes and their probabilities. Finally, traditional approaches may lack the agility and adaptability required to address today's rapidly changing risk landscape. Organizations need flexible, data-driven strategies that respond quickly to emerging risks and threats [49], [50].

8.5 UNDERSTANDING AI IN CONSTRUCTION SAFETY

8.5.1 Overview of Artificial Intelligence in Construction

Artificial intelligence (AI) is making significant strides in the construction industry, introducing innovative solutions to address traditional challenges and enhance project efficiency [11]. AI utilizes machine learning algorithms, data analytics, and computer vision to automate tasks, analyze data, and make informed decisions [51]. In the planning and design phase, AI can optimize building layouts, structural designs, and material selection based on historical project data, leading to more cost-effective and sustainable outcomes [9]. AI-powered project management systems assist in scheduling, resource allocation, and risk assessment, predicting potential delays and providing insights for better decision-making [52]. Construction robotics, enabled by AI, can perform repetitive and labor-intensive tasks, such as bricklaying and concrete pouring, increasing construction speed and accuracy [53]. Safety monitoring systems equipped with AI and computer vision can detect potential hazards in real time, reducing the risk of accidents [54]. AI also improves quality control by analyzing data from sensors and drones to identify defects and ensure compliance with standards.

Additionally, AI optimizes the supply chain by predicting material demands and managing inventory efficiently. Building management systems driven by AI maximize energy consumption, enhancing energy efficiency in construction projects. Predictive maintenance using AI algorithms analyzes equipment sensor data, predicting maintenance needs to avoid costly downtime. Moreover, autonomous vehicles and equipment, empowered by AI, improve productivity and safety on construction sites. Drones integrated with AI conduct site surveys, monitor progress, and capture high-resolution images for documentation and analysis, facilitating better decision-making. As AI technologies evolve, they hold immense promise in reshaping the construction industry, making it more efficient, safer, and sustainable [55], [56].

8.5.2 Machine Learning and Deep Learning Algorithms

In construction safety, machine learning and deep learning algorithms have emerged as powerful tools for improving site safety and mitigating potential risks [57]. These algorithms leverage data from various sources, such as sensors, cameras, and historical incident reports, to enhance safety protocols and decision-making processes [58]. Machine learning algorithms are applied to vast datasets of past safety incidents and near-miss reports to identify patterns and root causes of accidents [59]. By analyzing this data, these algorithms can predict potential safety hazards and suggest preventive measures to avoid similar incidents in the future [60]. This proactive approach enables construction companies to implement targeted safety measures, reducing the likelihood of accidents and injuries on construction sites.

Deep learning algorithms, specifically computer vision techniques, have revolutionized safety monitoring in construction. By processing real-time video feeds from cameras placed on construction sites, deep. Learning models can automatically detect and recognize safety hazards, such as workers not wearing personal protective equipment (PPE), unauthorized personnel in restricted areas, or potential equipment malfunctions. When a hazardous situation is identified, the system can alert relevant personnel immediately, enabling swift intervention and preventing potential accidents [61], [62].

Furthermore, these algorithms can be integrated with Internet of Things (IoT) sensors embedded in construction equipment and wearable devices worn by workers. The data collected from these sources is then processed using machine learning and deep learning algorithms to assess worker behavior, detect fatigue, and monitor equipment health. Construction companies can implement timely interventions and ensure a safer working environment by identifying unsafe conditions or risky behavior in real time. The continuous learning capability of these algorithms allows them to adapt and improve over time, leading to more accurate safety predictions and recommendations. Construction companies can use this information to develop targeted training programs, safety guidelines, and risk management strategies, leading to a safer work environment for all workers on-site [63], [64].

8.5.3 Data Collection and Analysis for AI Implementation

Data collection and analysis are essential for successful AI implementation in construction safety. Collecting relevant data is the foundation of any AI system to improve safety on construction sites. Data can be sourced from sensors, wearables, cameras, and other IoT devices deployed on the site [65]. These devices capture worker movements, equipment usage, environmental conditions, and potential hazards. Accurate and comprehensive data collection is crucial to developing AI models that effectively identify patterns, trends, and potential risks. For instance, data may include information about past accidents, near misses, safety inspection reports, and employee training records. Additionally, real-time data on worker behaviour, equipment status, and site conditions can be collected to support ongoing safety monitoring [66], [67].

Once the data is gathered, it undergoes thorough analysis to derive meaningful insights. Data analysis involves using statistical techniques, machine learning algorithms, and visualization tools to uncover patterns, correlations, and potential causes of safety incidents [68]. By analyzing historical data, AI models can learn from past accidents and near misses to identify common risk factors and develop predictive capabilities [69].

The analyzed data can be used to train AI models to detect and predict safety hazards on construction sites. For example, machine learning algorithms can recognize unsafe behaviors, such as not wearing proper personal protective equipment or working in dangerous areas. These AI models can also identify potential safety risks, such as hazardous site conditions or equipment malfunctions [9].

Moreover, AI models can continuously learn and adapt as new data is collected and analyzed. This iterative process ensures that the AI system remains up-to-date with the latest information and can respond effectively to evolving safety challenges [70]. Data privacy and security are critical considerations during data collection and analysis. Construction companies must ensure that sensitive information is protected and data is anonymized or aggregated appropriately to preserve worker privacy.

8.6 AI-DRIVEN RISK ASSESSMENT IN CONSTRUCTION

8.6.1 Predictive Analysis for Identifying Hazardous Areas

Predictive analysis is crucial in identifying hazardous areas in various contexts, including construction sites [71]. By analyzing historical data and real-time information, predictive models can effectively pinpoint regions more likely to be hazardous, allowing for proactive safety measures and risk mitigation strategies. In the construction industry, predictive analysis for identifying dangerous areas can be based on various data sources. This may include data from past incidents and near misses, safety inspection reports, worker behavior patterns, environmental conditions, and equipment status. By feeding this data into predictive models algorithms can identify patterns and correlations that might not be immediately apparent to human observers.

For example, predictive analysis can help construction companies identify areas with a higher frequency of accidents or near misses, indicating potential safety risks. It can also detect patterns in worker behavior, such as higher incident rates during specific times of the day or in certain weather conditions. Additionally, predictive models can consider the impact of factors like site layout, terrain, and the presence of heavy machinery on safety risks. By leveraging predictive analysis, construction companies can implement targeted safety measures in hazardous areas. For instance, additional safety barriers or warning signs can be placed in areas with a higher risk of accidents. Site managers can also schedule more frequent safety inspections in these zones to identify and address potential issues [55] proactively.

Furthermore, predictive analysis is not limited to static hazard identification [72]. It can continuously learn from new data, allowing the model to adapt and improve. As new data is collected and analyzed, the predictive model can refine its assessments, helping construction companies avoid potential safety hazards and reduce the likelihood of accidents. However, it is essential to remember that predictive analysis is not a substitute for human expertise and judgment. The insights provided by predictive models should be combined with the experience and knowledge of safety professionals to make well-informed decisions. Additionally, data quality and accuracy are critical for the success of

FIGURE 8.4 AI-Powered Detection System Enhances Construction Site Safety.

predictive analysis; therefore, data collection and validation processes must be robust and reliable [73]. Figure 8.4 shows the Human Element: Combining AI and Workforce Safety.

8.6.2 Real-Time Monitoring of Safety Parameters

Real-time monitoring of safety parameters is critical to ensuring a safe working environment in various industries, including construction, manufacturing, and transportation [74]. Organizations can promptly identify potential hazards, respond to emergencies, and implement proactive safety measures by continuously tracking and analyzing safety-related data in real time. Here are some key aspects and benefits of real-time safety monitoring.

8.6.2.1 Sensor Integration

Real-time safety monitoring relies on integrating various sensors and IoT devices. These sensors can be deployed on equipment, machinery, and vehicles and worn by workers as wearables. They collect data on temperature, humidity, air quality, vibration, noise levels, and worker vital signs [75].

8.6.2.2 Early Hazard Detection

Real-time monitoring allows for the early detection of safety hazards. When a safety parameter deviates from the acceptable range, the system can trigger immediate alerts to supervisors and workers, enabling them to take necessary actions promptly [76].

8.6.2.3 Emergency Response

In an emergency, real-time safety monitoring can automatically trigger response protocols, such as shutting down equipment or initiating evacuation procedures. This rapid response can prevent accidents or minimize their impact [77].

8.6.2.4 Predictive Analysis

Real-time data can be combined with historical data to perform predictive analysis. Machine learning algorithms can identify patterns and trends to predict potential safety risks, helping organizations proactively address them before they escalate [78].

8.6.2.5 Improved Worker Safety

Real-time safety monitoring prioritizes worker safety by providing instant feedback and alerts. Workers can be immediately notified of hazardous conditions or unsafe behaviors, encouraging them to take corrective actions and follow safety protocols [79].

8.6.2.6 Performance Tracking

Real-time monitoring allows organizations to track safety performance metrics continuously. By analyzing this data, they can identify trends and areas for improvement, leading to better safety practices and processes [80].

8.6.2.7 Compliance and Reporting

Real-time safety monitoring provides accurate and real-time data that can be used for compliance reporting. Organizations can demonstrate their commitment to safety and meet regulatory requirements more efficiently [81].

8.6.2.8 Remote Monitoring

Real-time safety monitoring can be done remotely, allowing safety personnel and managers to monitor multiple sites simultaneously. This remote access enhances operational efficiency and facilitates quick decision-making [82], [83].

8.6.2.9 Proactive Safety Measures

With real-time data insights, organizations can implement proactive safety measures, such as targeted training programs, equipment maintenance, and hazard-specific safety protocols.

8.6.2.10 Cultural Shift

Real-time safety monitoring fosters a safety-first culture within the organization. By prioritizing safety and using advanced technology to monitor it, organizations demonstrate their commitment to the well-being of their workforce [84].

8.6.3 Modeling and Simulation for Risk Mitigation

Modeling and simulation play a vital role in risk mitigation across various industries, including construction, finance, healthcare, and aerospace [85]. By creating virtual representations of real-world scenarios and systems, organizations can assess potential risks, test various mitigation strategies, and make informed decisions to enhance overall resilience. In risk mitigation, modeling involves creating mathematical or computational models that mimic the behavior of complex systems. These models are designed to capture the interactions between different variables and elements within the system. For example, in construction, a modeling approach might simulate the structural behavior of a building under various loading conditions to identify potential weak points and vulnerabilities [86].

Conversely, simulation involves running these models through iterative processes to understand how the system behaves under different conditions and scenarios. Organizations can gain insights into the possible outcomes and consequences of various risk factors by adjusting input parameters and running multiple simulations. The benefits of modeling and simulation for risk mitigation are manifold. Firstly, they provide a safe and controlled environment to assess risks without needing real-world implementation, which can be expensive and risky. Secondly, they allow organizations to explore various scenarios, including rare and extreme events that might not be easily observed or captured in real-life data. This helps identify "black swan" events that could have severe consequences [87], [43].

Modeling and simulation also enable organizations to test and compare different risk mitigation strategies before implementing them in the real world. This iterative process helps refine and optimize risk mitigation plans, increasing the likelihood of successful outcomes while minimizing negative impacts. Furthermore, modeling and simulation provide valuable insights into the dynamic nature of risks, particularly in complex and interconnected systems [88]. They help organizations understand how threats propagate and interact with one another, allowing for more holistic risk management strategies. In addition, these tools foster collaboration and communication among stakeholders by providing a visual representation of risks and their potential consequences. This facilitates decision-making processes and ensures that risk mitigation efforts are well-informed and agreed upon by all parties. It is essential to acknowledge that modelling and simulation accuracy rely on the data quality and assumptions used to build the models. Therefore, careful validation and calibration of models with real-world data are crucial to ensure their reliability [89].

8.6.4 AI-Enhanced Safety Inspections

AI-enhanced safety inspections are a cutting-edge application of AI in various industries, including construction, manufacturing, and transportation. By leveraging AI technologies such as computer vision, machine learning, and data analytics, safety inspections can be more efficient, accurate, and proactive [90]. One of the primary applications of AI in safety inspections is computer vision. AI-powered cameras and sensors can be deployed on-site to monitor and analyze safety conditions continuously. Computer vision algorithms can detect unsafe practices, potential hazards, and non-compliance with safety protocols. For example, they can identify workers not wearing appropriate PPE, hazardous work practices, or the presence of obstacles in walkways and working areas. Machine learning algorithms can be trained on historical safety inspection data to recognize patterns and trends in safety violations and incidents [91]. By analyzing this data, AI models can predict potential safety risks, allowing organizations to implement preventive measures and prioritize inspection efforts in high-risk areas. AI can also streamline the inspection process by automating routine tasks. For instance, drones with AI-powered cameras can inspect large and complex sites, capturing high-resolution images and videos for analysis. This automation reduces the need for manual inspections, saves time, and allows safety inspectors to focus on more critical tasks. Furthermore, AI can enhance the analysis of inspection data by identifying correlations and root causes of safety issues. By processing large datasets, AI algorithms can uncover insights that might be challenging for human inspectors to identify, leading to more effective risk mitigation strategies [92], [93].

Real-time monitoring is another significant advantage of AI-enhanced safety inspections. AI systems can provide instant feedback to workers and supervisors, issuing alerts and warnings when safety violations occur. This immediate feedback helps to correct unsafe behaviors promptly, reducing the likelihood of accidents and injuries and also contributes to creating a safety culture within organizations. By promoting transparency and data-driven decision-making, AI-enhanced safety inspections encourage proactive safety measures and improve employee safety awareness [94]. While AI-enhanced safety inspections offer numerous benefits, there are challenges to address, such as data privacy, model accuracy, and ongoing maintenance. Data security and compliance with privacy regulations are critical, especially when using video and sensor data. Continuous monitoring and fine-tuning of AI models are essential to maintain accuracy and adapt to changing conditions [95], [67].

8.7 AI-ENABLED SAFETY MANAGEMENT SYSTEMS

8.7.1 AI-Based Decision Support Systems for Safety Planning

AI-based decision support systems (DSSs) are becoming increasingly prevalent in safety planning across various industries, including construction, manufacturing, healthcare, and transportation. These systems leverage AI technologies, such as machine learning, data analytics, and optimization algorithms, to assist safety planners in making informed decisions and enhancing safety measures [88].

8.7.1.1 Risk Assessment and Prediction

AI-based DSS can analyze historical safety data, incident reports, and near-miss records to assess potential risks and predict safety hazards. The system can pinpoint high-risk areas and activities by identifying patterns and trends, enabling safety planners to focus on critical aspects [96].

8.7.1.2 Real-Time Monitoring and Alerts

AI-powered DSS can continuously monitor safety parameters and real-time data from sensors and cameras on construction sites or manufacturing facilities. If any unsafe conditions or violations are detected, the system can alert safety personnel immediately, enabling swift intervention and risk mitigation [96].

8.7.1.3 Safety Compliance and Regulations

AI-based DSS can keep safety planners up-to-date with the latest safety regulations and compliance requirements. By analyzing and interpreting regulatory changes, the system ensures that safety plans and protocols adhere to industry standards and legal obligations [96].

8.7.1.4 Optimal Resource Allocation

AI algorithms can optimize resource allocation for safety planning by considering budget constraints, workforce availability, and risk priorities. This ensures that safety measures are cost-effective and strategically aligned with the organization's goals [96].

8.7.1.5 Scenario Analysis and Simulation

AI-powered DSS can conduct scenario analysis and simulations to assess the potential impact of safety measures and interventions. By running simulations, safety planners can evaluate different strategies and their outcomes before implementation, facilitating evidence-based decision-making [96].

8.7.1.6 Personalized Safety Recommendations

AI can provide personalized safety recommendations and training for workers based on their behaviors, experiences, and safety records. This tailored approach enhances safety awareness and encourages compliance with safety protocols [96].

8.7.1.7 Continuous Learning and Improvement

AI-based DSS continuously learns and adapts to new safety data and feedback. This iterative process improves the system's accuracy and effectiveness over time, leading to better safety planning and risk management [96].

8.7.1.8 Data-Driven Insights

AI-powered DSS can analyze large volumes of data to provide data-driven insights and reports for safety planners. This empowers them to identify emerging safety trends, root causes of incidents, and opportunities for improvement [96].

While AI-based decision support systems offer significant benefits for safety planning, their successful implementation requires careful consideration of data quality, model transparency, and human-machine collaboration. Ensuring the accuracy and reliability of the AI models, as well as fostering a culture that values human expertise and collaboration with AI, are essential to harness the full potential of these systems in safety planning [96].

8.7.2 Automated Incident Reporting and Analysis

Automated incident reporting and analysis is a transformative application of technology in various industries, particularly in workplace safety and risk management. Organizations can streamline reporting and analysis of safety incidents by leveraging automation, AI, and data analytics, leading to quicker response times, improved accuracy, and more effective risk mitigation strategies. In automated incident reporting, sensors, cameras, and wearable devices are deployed on-site to monitor safety parameters continuously. When an incident occurs, such as an accident or near miss, these devices can automatically trigger incident reports, capturing critical data in real-time. This instantaneous reporting eliminates the delays associated with manual reporting, allowing safety personnel to respond promptly and take necessary actions [97], [98].

Furthermore, automation ensures that incident reports are more comprehensive and accurate. The data collected from sensors and devices provide objective and unbiased information, reducing the potential for human error or omissions. This enhanced data accuracy improves the quality of incident analysis, leading to a better understanding of the underlying causes and contributing factors. Automated incident analysis involves using AI and data analytics to quickly process large volumes of incident data. AI algorithms can identify patterns and trends in the data, helping safety professionals pinpoint common risk factors and emerging safety trends. By analyzing incident data in real time, organizations can gain valuable insights that can inform risk mitigation strategies and safety protocols [99].

Another significant advantage of automated incident reporting and analysis is performing root cause analysis more efficiently. AI algorithms can identify the primary causes of incidents by analyzing multiple incident reports and identifying commonalities among them. This enables safety personnel to address underlying issues and implement targeted safety measures to prevent similar incidents in the future. Additionally, automation allows for integrating incident data with other safety-related data, such as inspection reports, safety training records, and compliance data. This comprehensive data integration provides a holistic view of safety performance, enabling organizations to make data-driven decisions and allocate resources more effectively [100]. Overall, automated incident reporting and analysis significantly enhance the speed, accuracy, and effectiveness of safety incident management. By leveraging technology to automate the reporting process and analyze incident data in real-time, organizations can respond swiftly to incidents, identify underlying causes, and implement proactive safety measures to create safer and more secure working environments [101].

8.7.3 AI-Powered Safety Training and Education

AI-powered safety training and education are revolutionizing how organizations impart safety knowledge and skills to their workforce. By integrating AI into training programs, companies can create personalized, interactive, and dynamic learning experiences that are more engaging, effective, and adaptive to individual needs [102]. Figure 8.5 shows the Landscape of Machine Learning, Deep Learning, and Artificial Intelligence.

8.7.3.1 Personalized Learning Paths

AI algorithms can assess each employee's knowledge level and learning preferences and create personalized learning paths. This ensures that safety training is tailored to individual workers' specific needs and skill gaps, maximizing the movement's effectiveness [103].

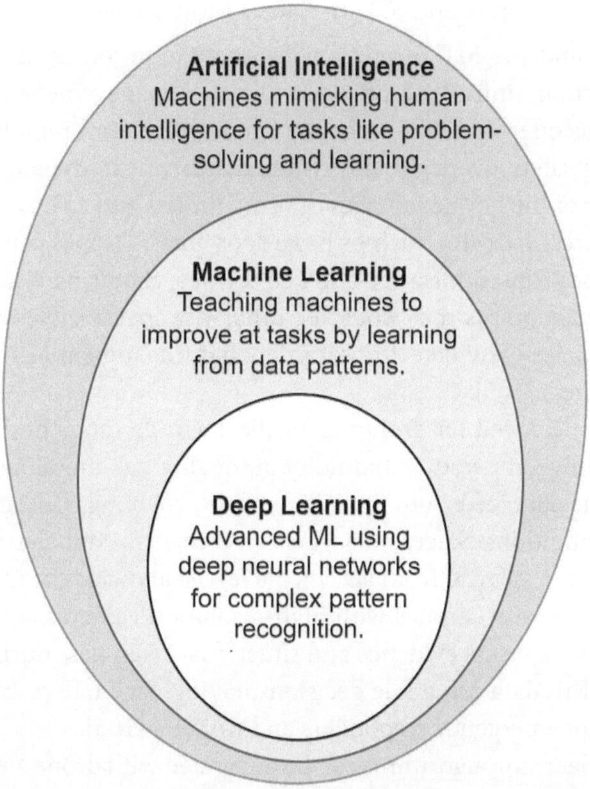

FIGURE 8.5 Illuminating the Landscape of Machine Learning, Deep Learning, and Artificial Intelligence.

8.7.3.2 Adaptive Learning

AI-powered training platforms can adapt the difficulty and content of training modules based on the learner's progress. This ensures that employees receive challenging and relevant content, promoting continuous learning and skill development [103].

8.7.3.3 Gamification and Interactivity

AI can infuse safety training with gamified elements, making learning more enjoyable and motivating. Interactive simulations, quizzes, and challenges can reinforce safety concepts and encourage active participation [103], [104].

8.7.3.4 Virtual Reality (VR) and Augmented Reality (AR)

AI-powered VR and AR technologies can provide immersive training experiences that simulate real-life safety scenarios. Employees can practice safety procedures in a safe virtual environment, preparing them for potential risks in the real world [103].

8.7.3.5 Continuous Monitoring and Feedback

AI algorithms can monitor employee performance during training, providing real-time feedback and progress tracking. This feedback loop helps employees identify areas for improvement and encourages continuous learning [103].

8.7.4 Integration of Drones and IoT for Safety Monitoring

Integrating drones and the IoT revolutionizes safety monitoring in various industries, including construction, infrastructure, and disaster management. Drones with IoT sensors and cameras offer mobility, real-time data capture, and remote monitoring capabilities, enhancing safety oversight and risk management in dynamic and challenging environments. One of the key advantages of using drones and IoT for safety monitoring is their ability to access hard-to-reach or hazardous areas. Drones can fly over construction sites, remote locations, or disaster-stricken regions, capturing visual data and sensor readings from vantage points that would be otherwise inaccessible or risky for human inspectors. This enables early detection of safety hazards, potential structural issues, or environmental dangers [105].

The IoT sensors deployed on drones can collect a wide range of data through wearable devices, including temperature, humidity, air quality, gas emissions, and even worker vitals. This real-time data feeds into the IoT network, enabling constant monitoring and analysis of safety conditions. Safety managers can receive instant alerts and notifications when safety parameters deviate from acceptable levels, allowing for immediate response and intervention. Moreover, drones with high-resolution cameras and thermal imaging capabilities can provide visual evidence and situational awareness during emergencies or hazardous events. This data can guide decision-making, facilitate rescue operations, and enhance the safety of emergency responders and workers [106].

AI and machine learning algorithms can be integrated with drone and IoT data for predictive safety analysis. By processing historical and real-time data, these algorithms can identify patterns, detect potential safety risks, and predict potential incidents. This proactive approach allows safety managers to implement preventive measures and allocate resources more effectively. The integration of drones and IoT for safety monitoring also streamlines inspection processes. Drones can automate aerial infrastructure inspections, such as bridges and power lines, reducing inspection time and costs.

Additionally, the data collected by drones can be fused with Building Information Modeling (BIM) and Geographic Information System (GIS) data, providing a comprehensive understanding of safety conditions and infrastructure health [107]. Despite the many benefits, data privacy, security, and regulatory compliance challenges need to be addressed. Ensuring the secure transmission and storage of sensitive data collected by drones and IoT sensors is crucial to protecting worker privacy and preventing data breaches [108].

8.8 FUTURE PERSPECTIVES

8.8.1 Potential Advancements in AI for Construction Safety

The future perspective of AI for construction safety holds immense potential for groundbreaking advancements that can revolutionize safety practices and foster a safer working environment. As AI technologies evolve, we can expect the following potential developments in construction safety [109].

8.8.1.1 Enhanced Real-Time Monitoring

AI-driven sensors and cameras will become even more sophisticated, enabling real-time monitoring of safety conditions on construction sites. These advanced sensors can detect a broader range of safety parameters, such as worker posture, equipment vibrations, and material handling, allowing for more comprehensive safety oversight. Predictive Analytics: AI will play a more significant role in predictive analytics, analyzing vast amounts of historical and real-time safety data to forecast potential safety risks.

Construction companies can proactively address safety concerns before incidents occur, leading to more effective risk mitigation strategies [110], [111].

8.8.1.2 AI-Driven Autonomous Vehicles and Robotics

Autonomous vehicles and robotics powered by AI can take over hazardous tasks, reducing the need for human workers to engage in high-risk activities. For instance, AI-driven drones and robots can be deployed for inspections, surveillance, and material handling, minimizing worker exposure to dangerous environments [8], [106].

8.8.1.3 Virtual Safety Training and Simulation

AI-powered virtual and augmented reality will enable immersive and realistic safety training experiences. Workers can practice safety procedures in virtual environments, gaining hands-on experience handling hazardous situations without real-world risks [106].

8.8.1.4 Wearable AI Devices

Wearable devices with embedded AI capabilities provide workers with personalized safety alerts and feedback. AI-driven wearables can monitor biometric data and detect signs of fatigue or stress, helping prevent accidents caused by worker fatigue [106].

8.8.2 The Role of AI in Promoting a Safety Culture

In the future, AI is set to play a pivotal role in promoting a safety culture across industries. As AI technologies advance, they will offer new and innovative ways to enhance safety practices, empower employees, and create a stronger safety-oriented mindset within organizations. One of the critical contributions of AI in promoting a safety culture is its ability to provide real-time insights and feedback. AI-powered sensors, wearables, and monitoring systems will continuously monitor safety parameters and worker behavior, instantly alerting employees and management to potential risks. This instant feedback reinforces safety protocols and encourages workers to be more vigilant about their safety practices [112].

AI can also facilitate personalized safety training and education. AI can tailor training programs to address specific skill gaps and safety needs by analyzing individual learning patterns and safety performance. This personalized approach will engage employees effectively, making safety training more relevant and meaningful. Furthermore, AI's predictive capabilities will enable organizations to proactively identify and address potential safety risks before they escalate. By analyzing vast amounts of data, AI algorithms can identify patterns and trends that might be missed by human observers, helping safety managers develop targeted risk mitigation strategies [113]. As AI technologies become more integrated with everyday operations, human-machine collaboration will become the norm. AI will not replace human expertise; instead, it will complement it. Workers will collaborate with AI-powered tools and systems to make better-informed decisions, making safety a collective responsibility, which will also support safety data analytics on a large scale. By processing massive datasets from various sources, AI will provide valuable insights into safety performance, incident trends, and areas for improvement. This data-driven approach will empower safety managers to make evidence-based decisions and allocate resources more efficiently [114]. Moreover, AI can foster a safety culture beyond organizational boundaries. As safety data is shared across industries and companies, best practices and safety lessons can be disseminated more effectively, leading to an industry-wide improvement in safety standards [115].

8.9 CONCLUSION

In conclusion, integrating AI technologies in various aspects of safety management and operations holds tremendous promise for enhancing safety practices across industries. AI's ability to process vast amounts of data, analyze complex patterns, and provide real-time insights enables organizations to proactively.

Identify and address safety risks, prevent accidents, and foster a safety culture. These applications offer unprecedented opportunities to create safer and more secure working environments, from AI-powered safety inspections and predictive analytics to AI-driven safety training and real-time monitoring. AI's continuous learning and adaptability ensure that safety practices remain up-to-date and effective, even in the face of evolving challenges.

As AI technologies advance and become more integrated with everyday operations, the future of safety management will be characterized by human-machine collaboration. AI will complement human expertise, empowering workers and safety managers to make more informed decisions and take proactive measures to protect employee well-being. However, alongside these transformative benefits, addressing challenges related to data privacy, security, and the responsible use of AI is essential. Ensuring transparency, ethics, and regulation compliance will be critical to building trust and confidence in AI-powered safety systems. Overall, the future of AI in safety management is promising. By embracing AI technologies, organizations can significantly reduce the likelihood of accidents, injuries, and safety incidents, creating workplaces that prioritize their workforce's health and safety. As AI continues to evolve, it will undoubtedly play an increasingly role in promoting a safety culture and transforming safety practices for the betterment of all.

REFERENCES

[1] D. Dodman, H. Leck, M. Rusca, and S. Colenbrander, "African urbanisation and urbanism: Implications for risk accumulation and reduction," *Int. J. Disaster Risk Reduct.*, vol. 26, pp. 7–15, Dec. 2017, doi: 10.1016/j.ijdrr.2017.06.029.

[2] R. A. Haslam, S. A. Hide, A. G. Gibb, D. E. Gyi, T. Pavitt, S. Atkinson, & A. R. Duff, "Contributing factors in construction accidents," *Appl. Ergon.*, vol. 36, no. 4, pp. 401–415, Jul. 2005, doi: 10.1016/j.apergo.2004.12.002.
[3] J. K. Wachter and P. L. Yorio, "A system of safety management practices and worker engagement for reducing and preventing accidents: An empirical and theoretical investigation," *Accid. Anal. Prev.*, vol. 68, pp. 117–130, Jul. 2014, doi: 10.1016/j.aap.2013.07.029.
[4] F. Zhang, A. P. C. Chan, A. Darko, Z. Chen, and D. Li, "Integrated applications of building information modeling and artificial intelligence techniques in the AEC/FM industry," *Autom. Constr.*, vol. 139, p. 104289, Jul. 2022, doi: 10.1016/j.autcon.2022.104289.
[5] J. Park, M. M. Salim, J. H. Jo, J. C. S. Sicato, S. Rathore, and J. H. Park, "CIoT-Net: a scalable cognitive IoT based smart city network architecture," *Human-centric Comput. Inf. Sci.*, vol. 9, no. 1, pp. 1–20, 2019.
[6] M. Parsamehr, U. S. Perera, T. C. Dodanwala, P. Perera, and R. Ruparathna, "A review of construction management challenges and BIM-based solutions: perspectives from the schedule, cost, quality, and safety management," *Asian J. Civ. Eng.*, vol. 24, no. 1, pp. 353–389, 2023.
[7] I. Okpala, C. Nnaji, and A. A. Karakhan, "Utilizing emerging technologies for construction safety risk mitigation," *Pract. Period. Struct. Des. Constr.*, vol. 25, no. 2, p. 4020002, 2020.
[8] S. O. Abioye, L. O. Oyedele, L. Akanbi, A. Ajayi, J. M. D. Delgado, M. Bilal, ... & A. Ahmed, "Artificial intelligence in the construction industry: A review of present status, opportunities and future challenges," *J. Build. Eng.*, vol. 44, p. 103299, 2021.
[9] Y. Pan and L. Zhang, "Roles of artificial intelligence in construction engineering and management: A critical review and future trends," *Autom. Constr.*, vol. 122, p. 103517, Feb. 2021, doi: 10.1016/j.autcon.2020.103517.
[10] N. Xu, L. Ma, L. Wang, Y. Deng, and G. Ni, "Extracting domain knowledge elements of construction safety management: Rule-based approach using Chinese natural language processing," *J. Manag. Eng.*, vol. 37, no. 2, p. 4021001, 2021.
[11] S. K. Baduge, S. Thilakarathna, J. S. Perera, M. Arashpour, P. Sharafi, B. Teodosio, ... & P. Mendis, *et al.*, "Artificial intelligence and smart vision for building and construction 4.0: Machine and deep learning methods and applications," *Autom. Constr.*, vol. 141, p. 104440, Sep. 2022, doi: 10.1016/j.autcon.2022.104440.
[12] M. H. Rahman, T. Islam, M. M. Rana, R. Tasnim, T. R. Mona, and M. M. Sakib, "Machine Learning Approach on Multiclass Classification of Internet Firewall Log Files," *arXiv Prepr. arXiv2306.07997*, 2023.
[13] G. Liu, M. Boyd, M. Yu, S. Z. Halim, and N. Quddus, "Identifying causality and contributory factors of pipeline incidents by employing natural language processing and text mining techniques," *Process Saf. Environ. Prot.*, vol. 152, pp. 37–46, 2021.
[14] W. Fang, L. Ding, P. E. Love, H. Luo, H. Li, F. Peña-Mora, ... & C. Zhou, "Computer vision applications in construction safety assurance," *Autom. Constr.*, vol. 110, p. 103013, 2020.
[15] H. Yan, N. Yang, Y. Peng, and Y. Ren, "Data mining in the construction industry: Present status, opportunities, and future trends," *Autom. Constr.*, vol. 119, p. 103331, 2020.
[16] M. Usama, H. Gardezi, F. E. Jalal, M. A. Rehman, N. Javed, S. Janjua, & M. Iqbal, "Predictive modelling of compression strength of waste GP/FA blended expansive soils using multi-expression programming," *Constr. Build. Mater.*, vol. 392, p. 131956, 2023.
[17] I. Awolusi, C. Nnaji, E. Marks, and M. Hallowell, "Enhancing construction safety monitoring through the application of internet of things and wearable sensing devices: A review," in

ASCE International Conference on Computing in Civil Engineering 2019, American Society of Civil Engineers Reston, VA, 2019, pp. 530–538.

[18] E. Celik and M. Gul, "Hazard identification, risk assessment and control for dam construction safety using an integrated BWM and MARCOS approach under interval type-2 fuzzy sets environment," *Autom. Constr.*, vol. 127, p. 103699, 2021.

[19] S. A. El-Safty and M. A. Shenashen, "Advanced nanoscale build-up sensors for daily life monitoring of diabetics," *Adv. Mater. Interfaces*, vol. 7, no. 15, p. 2000153, 2020.

[20] A. Singh, "Application of mobile edge computing & deep learning in mobile blockchain for security and safety," in *2023 3rd International Conference on Advance Computing and Innovative Technologies in Engineering (ICACITE)*, IEEE, 2023, pp. 2034–2038.

[21] A. Asadzadeh, M. Arashpour, H. Li, T. Ngo, A. Bab-Hadiashar, and A. Rashidi, "Sensor-based safety management," *Autom. Constr.*, vol. 113, p. 103128, 2020.

[22] Y. Li, H. Wei, Z. Han, J. Huang, and W. Wang, "Deep learning-based safety helmet detection in engineering management based on convolutional neural networks," *Adv. Civ. Eng.*, vol. 2020, pp. 1–10, 2020.

[23] K. Danna, "Health and well-being in the workplace: a review and synthesis of the literature," *J. Manage.*, vol. 25, no. 3, pp. 357–384, 1999, doi: 10.1016/S0149-2063(99)00006-9.

[24] A. Pamidimukkala, S. Kermanshachi, and T. Jahan Nipa, "Impacts of COVID-19 on health and safety of workforce in construction industry," in *International Conference on Transportation and Development 2021*, 2021, pp. 418–430.

[25] E. U. Asogwa, J. Blavo, and E. O. Asogwa, "Management Of Hazards And Prevention Of Accidents And Injuries In Building Construction Using 'Hierarchy Of Controls And Safety Education," 2022.

[26] E. J. Mamman, N. O. Okigbo, and A. A. Oke, "Evaluation of Occupational Safety Risk Factors in Building Construction Projects," *J. Appl. Ecol. Environ. Des.*, vol. 22, no. 4, pp. 165–174, 2022.

[27] S. Jafarzadeh Ghoushchi, S. Shaffiee Haghshenas, A. Memarpour Ghiaci, G. Guido, and A. Vitale, "Road safety assessment and risks prioritization using an integrated SWARA and MARCOS approach under spherical fuzzy environment," *Neural Comput. Appl.*, vol. 35, no. 6, pp. 4549–4567, 2023.

[28] J. B. Sexton, K. C. Adair, J. Profit, J. Bae, K. J. Rehder, T. Gosselin, ... & A. Frankel, "Safety Culture and Workforce Well-Being Associations with Positive Leadership WalkRounds," *Jt. Comm. J. Qual. Patient Saf.*, vol. 47, no. 7, pp. 403–411, Jul. 2021, doi: 10.1016/j.jcjq.2021.04.001.

[29] R. M. Crossley, M. H. Elmagrhi, and C. G. Ntim, "Sustainability and legitimacy theory: The case of sustainable social and environmental practices of small and medium-sized enterprises," *Bus. Strateg. Environ.*, vol. 30, no. 8, pp. 3740–3762, 2021.

[30] T. Islam, A. Kundu, R. J. Lima, M. H. Hena, O. Sharif, A. Rahman, & M. Z. Hasan, "5 Review Analysis of Ride-Sharing Applications Using Machine Learning Approaches," *Comput. Stat. Methodol. Model. Artif. Intell.*, pp. 99–122, 2023. CRC Press

[31] Z. Gizaw, "Public health risks related to food safety issues in the food market: a systematic literature review," *Environ. Health Prev. Med.*, vol. 24, pp. 1–21, 2019.

[32] A. Asgary, A. I. Ozdemir, and H. Özyürek, "Small and medium enterprises and global risks: Evidence from manufacturing SMEs in Turkey," *Int. J. Disaster Risk Sci.*, vol. 11, pp. 59–73, 2020.

[33] K. Lin, H. Chen, C. Y. Xu, P. Yan, T. Lan, Z. Liu, & C. Dong, "Assessment of flash flood risk based on improved analytic hierarchy process method and integrated maximum likelihood clustering algorithm," *J. Hydrol.*, vol. 584, p. 124696, May 2020, doi: 10.1016/j.jhydrol.2020.124696.

[34] A. Khan, S. Sepasgozar, T. Liu, and R. Yu, "Integration of BIM and Immersive Technologies for AEC: A Scientometric-SWOT Analysis and Critical Content Review," *Buildings*, vol. 11, no. 3, p. 126, Mar. 2021, doi: 10.3390/buildings11030126.

[35] T. Bakeli and A. Alaoui Hafidi, "A Fault Tree Analysis (FTA) based approach for construction projects safety risk management," *Proc. C 5th NA Detroit*, 2020.

[36] J. H. Purba, D. T. Sony Tjahyani, S. Widodo, and A. S. Ekariansyah, "Fuzzy probability based event tree analysis for calculating core damage frequency in nuclear power plant probabilistic safety assessment," *Prog. Nucl. Energy*, vol. 125, p. 103376, Jul. 2020, doi: 10.1016/j.pnucene.2020.103376.

[37] K. R. Golwalkar and R. Kumar, "Hazid, Hazop, and Ensuring Safety," in *Practical Guidelines for the Chemical Industry: Operation, Processes, and Sustainability in Modern Facilities*, Springer, 2022, pp. 11–30.

[38] E. Tepeli, F. Taillandier, and D. Breysse, "Multidimensional modelling of complex and strategic construction projects for a more effective risk management," *Int. J. Constr. Manag.*, vol. 21, no. 12, pp. 1218–1239, Dec. 2021, doi: 10.1080/15623599.2019.1606493.

[39] B. Hassani and B. K. Hassani, *Scenario analysis in risk management*. Springer, 2016.

[40] T. Luo, C. Wu, and L. Duan, "Fishbone diagram and risk matrix analysis method and its application in safety assessment of natural gas spherical tank," *J. Clean. Prod.*, vol. 174, pp. 296–304, Feb. 2018, doi: 10.1016/j.jclepro.2017.10.334.

[41] C. Lissak, A. Bartsch, M. De Michele, C. Gomez, O. Maquaire, D. Raucoules, & T. Roulland, "Remote sensing for assessing landslides and associated hazards," *Surv. Geophys.*, vol. 41, pp. 1391–1435, 2020.

[42] W. N. Adger, S. Dessai, M. Goulden, M. Hulme, I. Lorenzoni, D. R. Nelson, ... & A. Wreford, "Are there social limits to adaptation to climate change?," *Clim. Change*, vol. 93, pp. 335–354, 2009.

[43] T. Islam, M. D. A. Hosen, A. Mony, M. D. T. Hasan, I. Jahan, and A. Kundu, "A proposed Bi-LSTM method to fake news detection," in *2022 International Conference for Advancement in Technology (ICONAT)*, IEEE, 2022, pp. 1–5.

[44] M. Ramkumar, T. Schoenherr, and M. Jenamani, "Risk assessment of outsourcing e-procurement services: integrating SWOT analysis with a modified ANP-based fuzzy inference system," *Prod. Plan. Control*, vol. 27, no. 14, pp. 1171–1190, 2016.

[45] A. Beneking, S. Ellenbeck, and A. Battaglini, "Renewable energy cooperation between the EU and North Africa: Findings of a SWOT analysis," *Int. J. Energy Sect. Manag.*, vol. 10, no. 3, pp. 312– 336, 2016.

[46] C. C. Bonik, F. Akter, M. H. Rashid, and A. Sattar, "A Convolutional Neural Network Based Potato Leaf Diseases Detection Using Sequential Model," in *2023 International Conference for Advancement in Technology (ICONAT)*, IEEE, 2023, pp. 1–6.

[47] A. Jaafari, "Management of risks, uncertainties and opportunities on projects: time for a fundamental shift," *Int. J. Proj. Manag.*, vol. 19, no. 2, pp. 89–101, Feb. 2001, doi: 10.1016/S0263-7863(99)00047-2.

[48] A. Spence, W. Poortinga, and N. Pidgeon, "The psychological distance of climate change," *Risk Anal. An Int. J.*, vol. 32, no. 6, pp. 957–972, 2012.

[49] E. Zio, "Challenges in the vulnerability and risk analysis of critical infrastructures," *Reliab. Eng. Syst. Saf.*, vol. 152, pp. 137–150, Aug. 2016, doi: 10.1016/j.ress.2016.02.009.
[50] E. Zarei, F. Khan, and R. Abbassi, "How to account artificial intelligence in human factor analysis of complex systems?," *Process Saf. Environ. Prot.*, 2023.
[51] I. H. Sarker, "Ai-based modeling: Techniques, applications and research issues towards automation, intelligent and smart systems," *SN Comput. Sci.*, vol. 3, no. 2, p. 158, 2022.
[52] M. Kumar, "Application of artificial intelligence in civil engineering projects," *Math. Stat. Eng. Appl.*, vol. 70, no. 1, pp. 660–667, Jan. 2021, doi: 10.17762/msea.v70i1.2521.
[53] H. Wu, H. Li, X. Fang, and X. Luo, "A survey on teaching workplace skills to construction robots," *Expert Syst. Appl.*, vol. 205, p. 117658, Nov. 2022, doi: 10.1016/j.eswa.2022.117658.
[54] C. Park, D. Lee, and N. Khan, "An analysis on safety risk judgment patterns towards computer vision based construction safety management," in *Creative Construction e-Conference 2020*, Budapest University of Technology and Economics, 2020, pp. 31–38.
[55] W. J. Lee, H. Wu, H. Yun, H. Kim, M. B. G. Jun, and J. W. Sutherland, "Predictive maintenance of machine tool systems using artificial intelligence techniques applied to machine condition data," *Procedia CIRP*, vol. 80, pp. 506–511, 2019, doi: 10.1016/j.procir.2018.12.019.
[56] A. Sawhney, M. Riley, J. Irizarry, and M. Riley, "Construction 4.0," In Sawhney, A., Riley, M., Irizarry, J., Eds, 2020 Feb 12.
[57] A. Ajayi, L. Oyedele, H. Owolabi, O. Akinade, M. Bilal, J. M. Davila Delgado & L. Akanbi, "Deep learning models for health and safety risk prediction in power infrastructure projects," *Risk Anal.*, vol. 40, no. 10, pp. 2019–2039, Oct. 2020, doi: 10.1111/risa.13425.
[58] S. Mukherjee, S. Gupta, O. Rawlley, & S. Jain, "Leveraging big data analytics in 5G-enabled IoT and industrial IoT for the development of sustainable smart cities," *Trans. Emerg. Telecommun. Technol.*, vol. 33, no. 12, p. e4618, 2022.
[59] R. K. Gangadhari, V. Khanzode, & S. Murthy, "Application of rough set theory and machine learning algorithms in predicting accident outcomes in the Indian petroleum industry," *Concurr. Comput. Pract. Exp.*, vol. 34, no. 26, p. e7277, 2022.
[60] S. Zhang, J. Teizer, J.-K. Lee, C. M. Eastman, & M. Venugopal, "Building information modeling (BIM) and safety: Automatic safety checking of construction models and schedules," *Autom. Constr.*, vol. 29, pp. 183–195, 2013.
[61] Y. M. Goh, H. Brown, & J. Spickett, "Applying systems thinking concepts in the analysis of major incidents and safety culture," *Saf. Sci.*, vol. 48, no. 3, pp. 302–309, 2010.
[62] M. Q. Huang, J. Ninić, & Q. B. Zhang, "BIM, machine learning and computer vision techniques in underground construction: Current status and future perspectives," *Tunn. Undergr. Sp. Technol.*, vol. 108, p. 103677, Feb. 2021, doi: 10.1016/j.tust.2020.103677.
[63] C. Zhou & L. Y. Ding, "Safety barrier warning system for underground construction sites using Internet-of-Things technologies," *Autom. Constr.*, vol. 83, pp. 372–389, Nov. 2017, doi: 10.1016/j.autcon.2017.07.005.
[64] R. M. Choudhry, D. Fang, & S. M. Ahmed, "Safety management in construction: Best practices in Hong Kong," *J. Prof. issues Eng. Educ. Pract.*, vol. 134, no. 1, pp. 20–32, 2008.
[65] D. Georgakopoulos & P. P. Jayaraman, "Internet of things: from internet scale sensing to smart services," *Computing*, vol. 98, pp. 1041–1058, 2016.

[66] N. Haider, M. Z. Baig, & M. Imran, "Artificial intelligence and machine learning in 5G network security: Opportunities, advantages, and future research trends," *arXiv Prepr. arXiv2007.04490*, 2020.

[67] I. U. Khan *et al.*, "RSSI-controlled long-range communication in secured IoT-enabled unmanned aerial vehicles," *Mob. Inf. Syst.*, vol. 2021, pp. 1–11, 2021.

[68] C. I. Noshi & J. J. Schubert, "The role of machine learning in drilling operations; a review," in *SPE Eastern Regional Meeting*, SPE, 2018, p. D043S005R006.

[69] Z. Zhang, C. Trivedi, & X. Liu, "Automated detection of grade-crossing-trespassing near misses based on computer vision analysis of surveillance video data," *Saf. Sci.*, vol. 110, pp. 276–285, Dec. 2018, doi: 10.1016/j.ssci.2017.11.023.

[70] Y. Lu, "Artificial intelligence: a survey on evolution, models, applications and future trends," *J. Manag. Anal.*, vol. 6, no. 1, pp. 1–29, 2019.

[71] T. N. Van & T. N. Quoc, "Research trends on machine learning in construction management: a scientometric analysis," *J. Appl. Sci. Technol. Trends*, vol. 2, no. 03, pp. 96–104, 2021.

[72] K. Heilbrun, K. Yasuhara, & S. Shah, "Violence risk assessment tools: Overview and critical analysis," *Handb. Violence Risk Assess.*, pp. 11–28, 2011.

[73] B. T. Hazen, C. A. Boone, J. D. Ezell, & L. A. Jones-Farmer, "Data quality for data science, predictive analytics, and big data in supply chain management: An introduction to the problem and suggestions for research and applications," *Int. J. Prod. Econ.*, vol. 154, pp. 72–80, Aug. 2014, doi: 10.1016/j.ijpe.2014.04.018.

[74] K. Subramanya, S. Kermanshachi, & R. K. Patel, "The future of highway and bridge construction: Digital project delivery using integrated advanced technologies," in *International Conference on Transportation and Development 2022*, 2022, pp. 14–25.

[75] W.-F. Cheung, T.-H. Lin, & Y.-C. Lin, "A real-time construction safety monitoring system for hazardous gas integrating wireless sensor network and building information modeling technologies," *Sensors*, vol. 18, no. 2, p. 436, Feb. 2018, doi: 10.3390/s18020436.

[76] P. Allasia, A. Manconi, D. Giordan, M. Baldo, & G. Lollino, "ADVICE: A new approach for near-real-time monitoring of surface displacements in landslide hazard scenarios," *Sensors*, vol. 13, no. 7, pp. 8285–8302, Jun. 2013, doi: 10.3390/s130708285.

[77] F. Mirahadi & B. Y. McCabe, "EvacuSafe: A real-time model for building evacuation based on Dijkstra's algorithm," *J. Build. Eng.*, vol. 34, p. 101687, Feb. 2021, doi: 10.1016/j.jobe.2020.101687.

[78] A. Akbar, A. Khan, F. Carrez, & K. Moessner, "Predictive analytics for complex IoT data streams," *IEEE Internet Things J.*, vol. 4, no. 5, pp. 1571–1582, Oct. 2017, doi: 10.1109/JIOT.2017.2712672.

[79] X. Luo, H. Li, T. Huang, & M. Skitmore, "Quantifying hazard exposure using real-time location data of construction workforce and equipment," *J. Constr. Eng. Manag.*, vol. 142, no. 8, Aug. 2016, doi: 10.1061/(ASCE)CO.1943-7862.0001139.

[80] S. Sinelnikov, J. Inouye, & S. Kerper, "Using leading indicators to measure occupational health and safety performance," *Saf. Sci.*, vol. 72, pp. 240–248, Feb. 2015, doi: 10.1016/j.ssci.2014.09.010.

[81] A. Barnett, M. Winning, S. Canaris, M. Cleary, A. Staib, & C. Sullivan, "Digital transformation of hospital quality and safety: Real-time data for real-time action," *Aust. Heal. Rev.*, vol. 43, no. 6, p. 656, 2019, doi: 10.1071/AH18125.

[82] J. Park, K. Kim, & Y. K. Cho, "Framework of automated construction-safety monitoring using cloud-enabled BIM and BLE mobile tracking sensors," *J. Constr. Eng. Manag.*, vol. 143, no. 2, Feb. 2017, doi: 10.1061/(ASCE)CO.1943-7862.0001223.

[83] I. U. Khan, A. Abdollahi, R. Alturki, M. D. Alshehri, M. A. Ikram, H. J. Alyamani, & S. Khan, "Intelligent detection system enabled attack probability using Markov chain in aerial networks," *Wirel. Commun. Mob. Comput.*, vol. 2021, pp. 1–9, 2021.

[84] P. X. Zou, "Fostering a strong construction safety culture," *Leadersh. Manag. Eng.*, vol. 11, no. 1, pp. 11–22, Jan. 2011, doi: 10.1061/(ASCE)LM.1943-5630.0000093.

[85] A. Virtue, T. Chaussalet, and J. Kelly, "Healthcare planning and its potential role increasing operational efficiency in the health sector," *J. Enterp. Inf. Manag.*, vol. 26, no. 1/2, pp. 8–20, Feb. 2013, doi: 10.1108/17410391311289523.

[86] D. J. Wagg, K. Worden, R. J. Barthorpe, and P. Gardner, "Digital twins: state-of-the-art and future directions for modeling and simulation in engineering dynamics applications," *ASCE-ASME J. Risk Uncertain. Eng. Syst. Part B Mech. Eng.*, vol. 6, no. 3, p. 30901, 2020.

[87] K. O. Yoro, M. O. Daramola, P. T. Sekoai, U. N. Wilson, and O. Eterigho-Ikelegbe, "Update on current approaches, challenges, and prospects of modeling and simulation in renewable and sustainable energy systems," *Renew. Sustain. Energy Rev.*, vol. 150, p. 111506, 2021.

[88] S. Elhouar, E. Hochscheid, M. A. Alzarrad, and C. Emanuels, "Will artificial intelligence (AI) take over the construction world?-A multidisciplinary exploration," in *Creative construction e- conference 2020*, Budapest University of Technology and Economics, 2020, pp. 59–66.

[89] H. Jia, F. Chen, D. Pan, E. Du, L. Wang, N. Wang, & A. Yang, "Flood risk management in the Yangtze River basin—Comparison of 1998 and 2020 events," *Int. J. Disaster Risk Reduct.*, vol. 68, p. 102724, 2022.

[90] F. Ansari & L. Kohl, "Ai-enhanced maintenance for building resilience and viability in supply chains," in *Supply Network Dynamics and Control*, Springer, 2022, pp. 163–185.

[91] M. A. Khan, B. A. Alvi, A. Safi, & I. U. Khan, "Drones for good in smart cities: a review," in *Proc. Int. Conf. Elect., Electron., Comput., Commun., Mech. Comput.(EECCMC)*, 2018, pp. 1–6.

[92] I. Yitmen, I. Kovacic, & L. C. Tagliabue, "Cognitive digital twins for facilitating construction 4.0: Challenges and opportunities for implementation," *Front. Built Environ.*, vol. 9, p. 1130115, 2023.

[93] I. U. Khan, S. Z. Nain Zukhraf, A. Abdollahi, S. A. Imran, Qureshi, I. M., Aziz, M. A., & S. B. Hussian Shah, "Reinforce based optimization in wireless communication technologies and routing techniques using internet of flying vehicles," in *The 4th International conference on future networks and distributed systems (ICFNDS)*, 2020, pp. 1–6.

[94] M. Schlüter, H. Lickert, K. Schweitzer, P. Bilge, C. Briese, F. Dietrich, & J. Krüger, "AI-enhanced identification, inspection and sorting for reverse logistics in remanufacturing," *Procedia CIRP*, vol. 98, pp. 300–305, 2021.

[95] S. Gul, M. El-Zein, H. Aldin, M. Aldin, C. Mullin, and E. van Oort, "Automated mud check with an AI-enhanced automated mud skid: results of a long-term permian field trial," in *SPE/IADC Drilling Conference and Exhibition*, SPE, 2022, p. D031S022R002.

[96] C. J. Smith and A. T. C. Wong, "Advancements in artificial intelligence-based decision support systems for improving construction project sustainability: a systematic literature review," in *Informatics*, MDPI, 2022, p. 43.

[97] V. Singhal, S. S. Jain, D. Anand, A. Singh, S. Verma, J. J. Rodrigues, ... & C. Iwendi, "Artificial intelligence enabled road vehicle-train collision risk assessment framework for unmanned railway level crossings," *IEEE Access*, vol. 8, pp. 113790–113806, 2020.

[98] I. U. Khan, I. M. Qureshi, M. A. Aziz, T. A. Cheema, and S. B. H. Shah, "Smart IoT control-based nature inspired energy efficient routing protocol for flying ad hoc network (FANET)," *IEEE Access*, vol. 8, pp. 56371–56378, 2020.

[99] S. Rao, "The benefits of AI in construction," *Retrieved March*, vol. 16, p. 2020, 2019.

[100] R. Sacks, M. Girolami, and I. Brilakis, "Building information modelling, artificial intelligence and construction tech," *Dev. Built Environ.*, vol. 4, p. 100011, 2020.

[101] Y. Cao, Y. Zandi, A. S. Agdas, Q. Wang, X. Qian, L. Fu, ... & A. Roco-Videla, "A review study of application of artificial intelligence in construction management and composite beams," *Steel and Composite Structures, An International Journal*, vol. 39, no. 6, pp. 685–700, 2021.

[102] R. Chakkravarthy, "Artificial intelligence for construction safety," *Prof. Saf.*, vol. 64, no. 1, p. 46, 2019.

[103] C.-H. Ko and M.-Y. Cheng, "Hybrid use of AI techniques in developing construction management tools," *Autom. Constr.*, vol. 12, no. 3, pp. 271–281, 2003.

[104] I. U. Khan, S. B. H. Shah, L. Wang, M. A. Aziz, T. Stephan, and N. Kumar, "Routing protocols & unmanned aerial vehicles autonomous localization in flying networks," *Int. J. Commun. Syst.*, p. e4885, 2021.

[105] J. B. H. Yap, M. Skitmore, C. G. Y. Lam, W. P. Lee, and Y. L. Lew, "Advanced technologies for enhanced construction safety management: investigating Malaysian perspectives," *Int. J. Constr. Manag.*, pp. 1–10, 2022.

[106] I. Jeelani and M. Gheisari, "Safety challenges of UAV integration in construction: Conceptual analysis and future research roadmap," *Saf. Sci.*, vol. 144, p. 105473, 2021.

[107] S. H. Alsamhi, O. Ma, M. S. Ansari, and F. A. Almalki, "Survey on collaborative smart drones and internet of things for improving smartness of smart cities," *Ieee Access*, vol. 7, pp. 128125–128152, 2019.

[108] A. Kumar and S. Jain, "Drone-based monitoring and redirecting system," *Dev. Futur. Internet Drones Insights, Trends Road Ahead*, pp. 163–183, 2021.

[109] V. Tsiatsis, S. Karnouskos, J. Holler, D. Boyle, and C. Mulligan, *Internet of Things: Technologies and Applications for a New Age of Intelligence*. Academic Press, 2018.

[110] M. L. Das, "Privacy and security challenges in internet of things," in *Distributed Computing and Internet Technology: 11th International Conference*, ICDCIT 2015, Bhubaneswar, India, February 5-8, 2015. Proceedings 11, Springer, 2015, pp. 33–48.

[111] I. U. Khan, A. Abdollahi, A. Jamil, B. Baig, M. A. Aziz, and F. Subhan, "A novel design of FANET routing protocol aided 5G communication using IoT," *J. Mob. Multimed.*, pp. 1333–1354, 2022.

[112] R. Jin, X. W. Zou Patrick, Poorang Piroozfar, Hannah Wood, Yang Yang, Libo Yan, and Yu Han. *et al.*, "A science mapping approach based review of construction safety research," *Saf. Sci.*, vol. 113, pp. 285–297, 2019.

[113] T. D. Akinosho, L. O. Oyedele, M. Bilal, A. O. Ajayi, M. D. Delgado, O. O. Akinade, & A. A. Ahmed, "Deep learning in the construction industry: A review of present status and future innovations," *J. Build. Eng.*, vol. 32, p. 101827, 2020.

[114] D. Hardison and M. Hallowell, "Construction hazard prevention through design: Review of perspectives, evidence, and future objective research agenda," *Saf. Sci.*, vol. 120, pp. 517–526, 2019.

[115] Y. Hooda, P. Kuhar, K. Sharma, and N. K. Verma, "Emerging applications of artificial intelligence in structural engineering and construction industry," in *Journal of Physics: Conference Series*, IOP Publishing, 2021, p. 12062.

CHAPTER 9

Efficient Energy Utilization in Smart Grids

An Artificial Intelligence Perspective

Zakria Qadir, Yasir Ali Khan,
Muhammad Tausif Afzal Rana, Fareed Ud Din,
and Niusha Shafiabady

9.1 INTRODUCTION

Efficient energy utilization in smart grids, especially in the context of Direct Current (DC) imbalance, is a critical issue that can be effectively addressed through an artificial intelligence perspective, which has been proven to be a revolutionizing factor via recent research in several domains [1]. In this context, this chapter delves into applying advanced artificial intelligence (AI) techniques to optimize energy management and distribution in smart grid systems. With the help of AI algorithms, this approach allows for real-time monitoring, prediction, and control of DC imbalances in the grid, enabling the efficient utilization of energy resources. By leveraging AI, smart grids can adapt to dynamic energy demands and prioritize renewable energy sources, minimizing waste and ensuring grid stability.

In this modern era, renewable energy sources are widely used to meet power and energy requirements at both commercial and non-commercial levels. The excess use of these renewable energy resources can lead to a shortage forever. Therefore, developing such a technology that can efficiently utilize these resources is critical. Mostly renewable energy sources (i.e., wind, solar, hydro) are used in the power sector either to power up electronic devices or modern cars at charging stations. Usually, hybrid schemes (i.e., more than one Renewable Energy Source (RES)) are preferred to improve power system reliability [2–3]. In hybrid power generation systems, renewable energy sources (RESs) such as wind and photovoltaic (PV) systems are commonly utilized due to their inherent intermittency and fluctuation characteristics. To address these challenges, energy storage units such as supercapacitors and superconducting material energy sources (SMEs) play a crucial role.

DOI: 10.1201/9781032648323-9

These storage units help stabilize the output from RESs, ensuring a more reliable and consistent power supply [6,23,27].. Both photovoltaic (PV) and wind energy technologies are well-established and can complement each other, offering flexibility in substituting one another depending on environmental conditions and energy demands [3].

Different hybrid power generation-related topics covering power smoothing schemes, controlling mechanisms, power management methods, and stability analysis have already been discussed in previous research [3], [4]. The output fluctuation of hybrid power generators is a real challenge to grid security such as voltage flicker, fluctuation, harmonics, etc. [4]. A detailed overview of power generation-related topics is covered in the coming sections. This research also covers power smoothing.

A hybrid power generation system, encompassing both PV and wind sources, offers three distinct configurations: DC coupled, AC coupled, and DC/AC coupled [5]. All RESs in the DC configuration are interconnected on the same DC bus using power electronics interfaces. DC loads receive direct power from the DC bus, and AC loads require a DC/AC inverter. Conversely, the AC configuration involves linking all RESs to a shared AC bus, with the PV array connected via a three-phase DC-AC inverter that converts the DC output voltage to a standard three-phase AC voltage. The DC/AC configuration combines characteristics from both the DC and AC setups.

Integrating renewable energy sources, such as photovoltaic (PV) and wind power, into grid-tied systems is instrumental in transitioning toward a more sustainable and environmentally conscious energy infrastructure [7]. Grid-tied hybrid PV-wind systems hold significant promise by harnessing the complementary characteristics of both energy sources to ensure a consistent power supply [8]. However, one of the significant challenges in these systems is the occurrence of DC-link imbalances, which can lead to inefficiencies, power losses, and grid instability. To address this critical issue, researchers have turned to applying AI [9]. This chapter delves into the extensive knowledge surrounding the optimization of DC-link imbalance in grid-tied hybrid PV-wind systems from an AI perspective. By harnessing AI algorithms and data-driven approaches, this research strives to develop solutions that enhance energy yield and ensure the seamless integration of renewable energy into the existing grid infrastructure. In this chapter, we propose a system to deal with fluctuations and to provide smooth AC voltage to the load. We also present the integration of a hybrid power generation system with the grid to justify the system's performance. The complete system is developed using the MATLAB® tool.

This chapter is organized as follows. In Section 9.2, an overview of the previous research made on power generation-related topics is presented. The design methodology follows the literature review. Section 9.3 describes the design methodology and all the approaches used to achieve robust results for power smoothing. In Section 9.4, results and discussion are presented. This section shows the novelty of the proposed system. Section 9.5 provides the concluding remarks.

9.2 LITERATURE STUDY

Recent studies cover a diverse range of research papers and scholarly articles delving into the integration of AI techniques within the realm of smart grids and DC imbalance mitigation. These studies shed light on different methodologies and AI algorithms utilized to tackle DC imbalances, underscoring their capacity to boost grid efficiency and sustainability. Additionally, they delve into the challenges and limitations faced by researchers in this domain, pinpointing potential paths for future exploration. The related study section establishes a groundwork for comprehending and valuing the field's evolution, emphasizing the significance of ongoing research and innovation in this domain.

Sneha et al. [10] introduced a hybrid energy management system comprising a PV array, a polymer electrolyte membrane fuel cell, and a Lithium-ion (Li-ion) battery serving as auxiliary energy storage. DC-DC converters establish the interface among the PV-FC-battery combination. Energy management among these three sources is based on the load's power requirements. The PV array operates in either maximum power point tracking (MPPT) or desired power control (DPC) mode. The Li-ion battery concurrently operates in voltage control mode, maintaining its state of charge (SOC) and the DC link. The fuel cell (FC) functions in power control modes (PCMs) using regulated DC-link voltage to meet active power demands. One limitation of DPC is that, under maximum SOC conditions, the excess power generated by the PV is not utilized. The hybridization of the fuel cell addresses the drawbacks of relying on a single PV source. The proposed system underwent simulation using MATLAB/Simulink®.

Yaow et al. [11] introduced an enhanced hybrid PV/Wind power generation system utilizing a multi-input converter with a DC bus voltage regulation strategy. The system comprises a multi-input DC/DC converter and a full-bridge inverter. A DC-DC converter is employed to extract maximum power from the PV array and wind turbine. The full-bridge inverter injects nearly sinusoidal AC current into the utility grid to maintain a balanced input-output power flow. The suggested DC bus voltage regulation method can manage the variation in DC bus voltage derived from either the PV array or the wind turbine, as well as the size of the DC bus capacitor bank. Additionally, the change in the amplitude of the injected AC current remains moderate, leading to an improvement in the power quality of the utility. The proposed DC bus regulation method underwent testing on a 1.2KW prototype system for feasibility validation.

Hongjie et al. [12] introduced a three-terminal AC/DC hybrid microgrid featuring two DC terminals and one AC terminal. The system comprises an AC grid interface based on cascaded H-bridge (CHB) converters and a DC sub-grid interface with two dual active bridge (DAB) converters connecting isolated DC buses. The DC rails of the CHB converters are directly connected to the DAB converters to minimize the number of power conversion stages. The proposal also includes injecting zero-sequence voltage in the CHB converters to address the issues of imbalanced grid currents and DC rail voltage. The study delves into investigating the impacts of control parameters on the system's stability and dynamic response.

Li et al. [13] presented a method to enhance the stability of a multi-machine power system integrated with a hybrid power generation system involving wind and photovoltaic sources, utilizing energy storage based on supercapacitors. The functionality of the proposed system was simulated using a 300-MW wind turbine generator and an equivalent 75 MW PV array. The PV and wind power systems are linked to the DC link through a voltage-source converter and a DC/DC boost converter. The DC-link's power is then transmitted to the multi-machine system through a source inverter, step-up transformers, and a connecting line. To address power fluctuations resulting from variations in wind speed or solar irradiance, a supercapacitor-based energy storage unit is connected to the shared DC link via a bidirectional DC/DC converter for smoothing purposes. Furthermore, a proportional-integral-derivative (PID)-supplementary damping controller (PID-SDC) was proposed for the bidirectional DC/DC converter of the supercapacitor to enhance the damping characteristics of the low-frequency oscillations associated with the studied multi-machine power system.

Lalit et al. [14] introduced an innovative standalone system incorporating solar, wind, a storage battery, and a diesel generator, distinguished by using PLL control and dump power control. The system's performance was evaluated using MATLAB/Simulink. Lalit et al. also explored contemporary power control techniques for various sources, including solar, wind, diesel generators, and storage batteries. The dump power control facilitates feedback control, eliminating the need for a separate control device. The proposed method demonstrates efficiency in terms of active, reactive, and dump power control. Experimental results suggest that this approach could be highly beneficial for integrating diverse power generation sources, as seen in related studies [15, 16].

Hu et al. [17] suggested a hybrid power generation system, combining offshore wind and tidal energy, to enhance the efficiency and reliability of power generation. Additionally, they introduced a distinctive control approach designed to mitigate power fluctuations. In this method, the tidal turbine serves to offset variations in wind power and a tidal power coefficient is introduced to quantify the compensation ratio of the tidal power generation system.

9.3 METHODOLOGY

The methodology employed in optimizing the DC-Link Imbalance in Grid-Tied Hybrid PV-Wind Systems from an AI perspective involves a multifaceted approach. To address the challenge of DC-link imbalances, researchers often begin by collecting real-world data from the hybrid PV-Wind system, including information on energy generation and consumption, environmental factors, and system components [17–20]. Machine learning and AI algorithms, such as neural networks, reinforcement learning, and predictive analytics, are then employed to analyze this data and develop intelligent models to predict and mitigate DC-link imbalances [20–22]. Additionally, advanced control strategies are implemented to dynamically adjust the PV and wind components' operation, optimizing their output in response to changing conditions [24–27]. However, in this chapter, we used the simulation and modeling tools to evaluate the proposed AI-based solutions under various scenarios, and performance metrics are used to assess the effectiveness of the optimization process.

9.3.1 Proposed Hybrid Model

An intrusion detection system is mostly used to identify security attacks. However, anomaly-based Intrusion detection system (IDS) formulate a threshold to detect unknown attacks. Data traffic monitoring is associated with packet headers.

The proposed hybrid model consists of a PV system, wind turbine, DC imbalance block, battery controller unit, and inverter control, as shown in Figure 9.1. The Canadian PV system (Solar CS6X-300M) comprises ten series-connected modules in a single parallel string. Every single module has a maximum power of 300.03 W, and a combination of 10 provides 3 kW of total power output. A DC-DC boost converter circuit combined with the solar panel elevates the output power of the solar system, and the P& O algorithm helps to reach the MPPT. To test the efficacy of the solar system, varying irradiance and temperature are applied to the solar panel, as shown in Figure 9.1. The P & O algorithm continuously compares the power vs. voltage curve. It maintains the maximum power to be fed to the load.

The horizontal wind turbine model (Eteno AN-3000) with a rated power of 3kW is proposed for the hybrid system. The wind subsystem comprises a wind generator equipped with a permanent magnet synchronous generator (PMSG) and integrated with a rectifier to convert AC to DC voltage level. The wind turbine produces a mechanical torque (Tm) depending directly on the generator speed, wind speed, and the pitch angle. Thus, the product of generated Tm and the base torque provide the necessary torque to drive a PMSG to produce AC electric power. A DC-DC buck converter circuit is also integrated to maintain the common DC-link (output DC voltage at the desired level).

9.3.2 DC Imbalance Signal

To test the efficacy of the proposed hybrid system, a DC imbalance signal is applied at the input of an inverter between the DC-link terminals. As a DC step input signal is applied to the inverter input, it results in a DC imbalance at 0.05 sec and can cause the overall system to malfunction, as shown in Figure 9.2.

In order to modify it, the inverter control having a PI controller provides a controlling mechanism to the phase and line voltage being delivered to the load. The inverter output waveform for the hybrid model with and without the Dc-link imbalance is shown in Figure 9.3. The one without a Dc-link imbalance, as illustrated in Figure 9.3(a), has a Kp of 20.6, and a Ki of 36 shows the desired output waveform to be fed to the load. While, with the Dc-link imbalance, the value of Kp is lessened to 0, keeping the ki to be the same (36), a more stabilized output voltage waveform can be seen for both the phase and line voltage after a delay of 0.1 sec, respectively, as shown in Figure 9.3(b).

9.3.3 Power Storage component

Nickel-metal hydride (Ni-MH) batteries have emerged as a prominent power storage component, offering a range of advantages in various applications. Ni-MH batteries are known for their high energy density, making them well-suited for consumer electronics. Their relatively low self-discharge rate and longer cycle life than many other battery types make them a reliable choice for long-term energy storage. Moreover, Ni-MH batteries are environmentally friendly, as they contain no toxic materials like cadmium and are easier to recycle. These attributes have played a pivotal role in their extensive adoption in renewable energy systems. The hybrid system incorporates a Ni-MH battery to store generated power. In instances where the hybrid system does not generate sufficient power, this stored energy is utilized to supply subsequent power to the load. The battery, with a

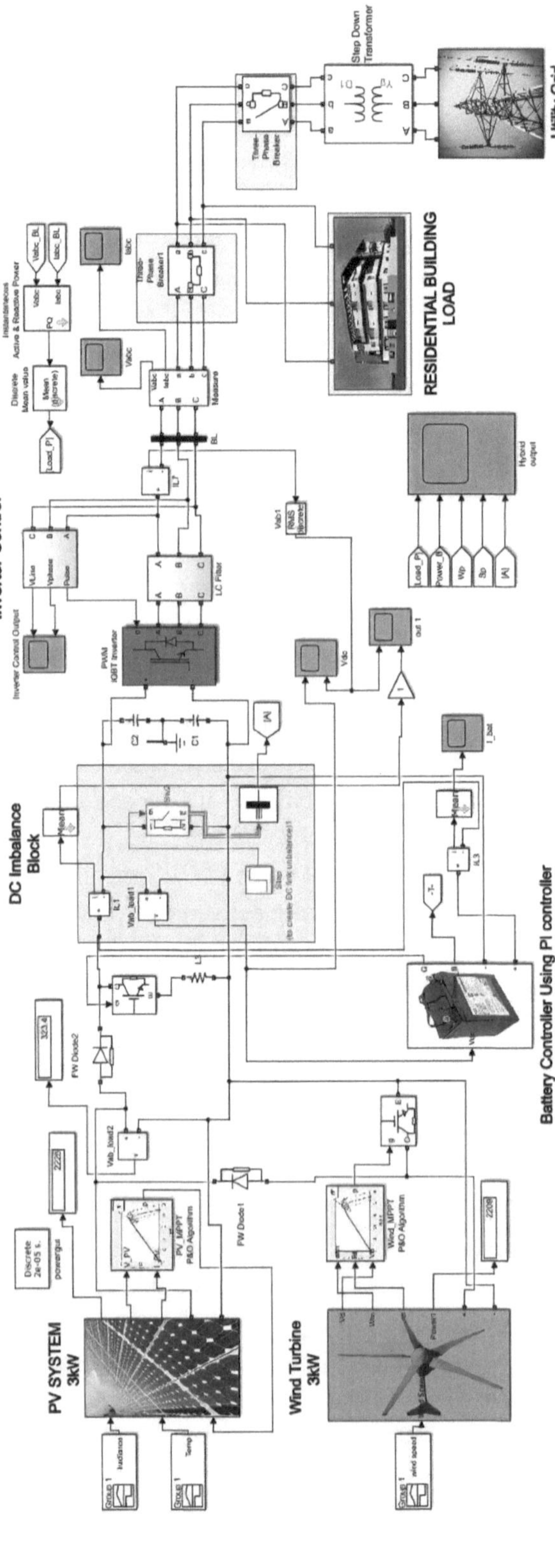

FIGURE 9.1 System design of the hybrid system. Zoom in for a better view.

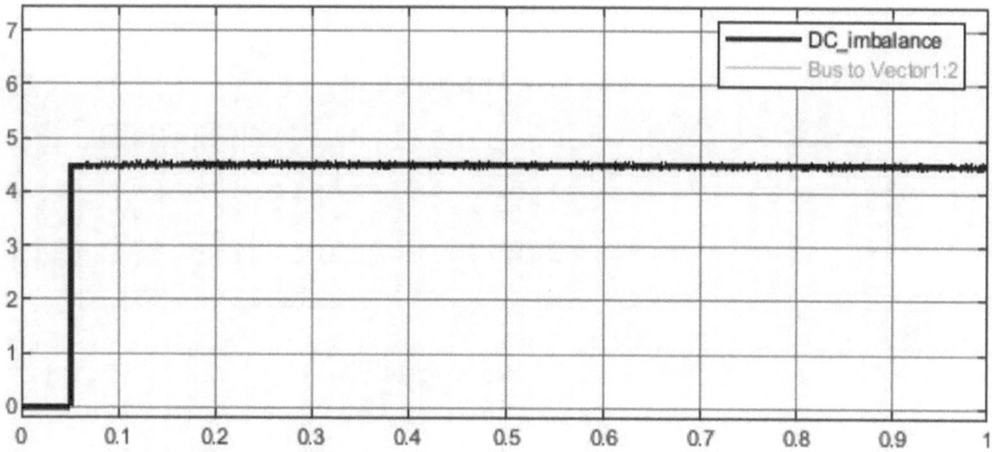

FIGURE 9.2 DC Imbalance signal.

capacity of 1.3 kWh (200 V, 6.5 Ah), can sustain the load with energy for approximately an hour (60 min), as illustrated in Figure 9.4.

Though the battery size can be increased to supply power to load for the desired time period, as this hybrid system is grid-tied, the excess power stored in the battery can be shared with the grid. This efficient net metering technique will allow the consumer to get incentive over the power they have saved, resulting in much economic electricity production and also free of GHG emissions (advancing renewable energy technology to be used rather than depending on fossil fuels).

The battery controller model consists of a closed-loop circuit and two subsystems. The first subsystem is related to the Ni-MH battery storage system integrated with a DC-DC converter, while the second subsystem is entirely the battery controller.

9.3.3.1 Battery Controller

The output hybrid voltage (Vdc) is compared with the reference voltage and fed to the PI controller, which stabilizes the signal, as shown in Figure 9.5. Furthermore, a data type converter is employed to ensure that real-time values of the input and output are consistent. This results in generating two gate signals (S1, S2) for the Insulated Gate Bipolar Transistor (IGBT) in the battery storage system for DC-DC conversion. Specifically, S1 represents the gate pulse when the battery is fully charged, while S2 signifies that the battery has fallen below a certain level (discharged).

Additionally, to synchronize the hybrid output voltage with the state of charge (SOC) of the battery, both of these signals(voltages) are compared using a PI controller, and based on the SOC to be greater than 80%, a gate signal is generated which is fed to the IGBT at the output of the hybrid system to validate that the hybrid system has generated the power, so no need of battery output is required. These signals (S1, S2, and Gate), determined based on the Vdc, SOC, and I_bat, are fed back to them with the input voltage.

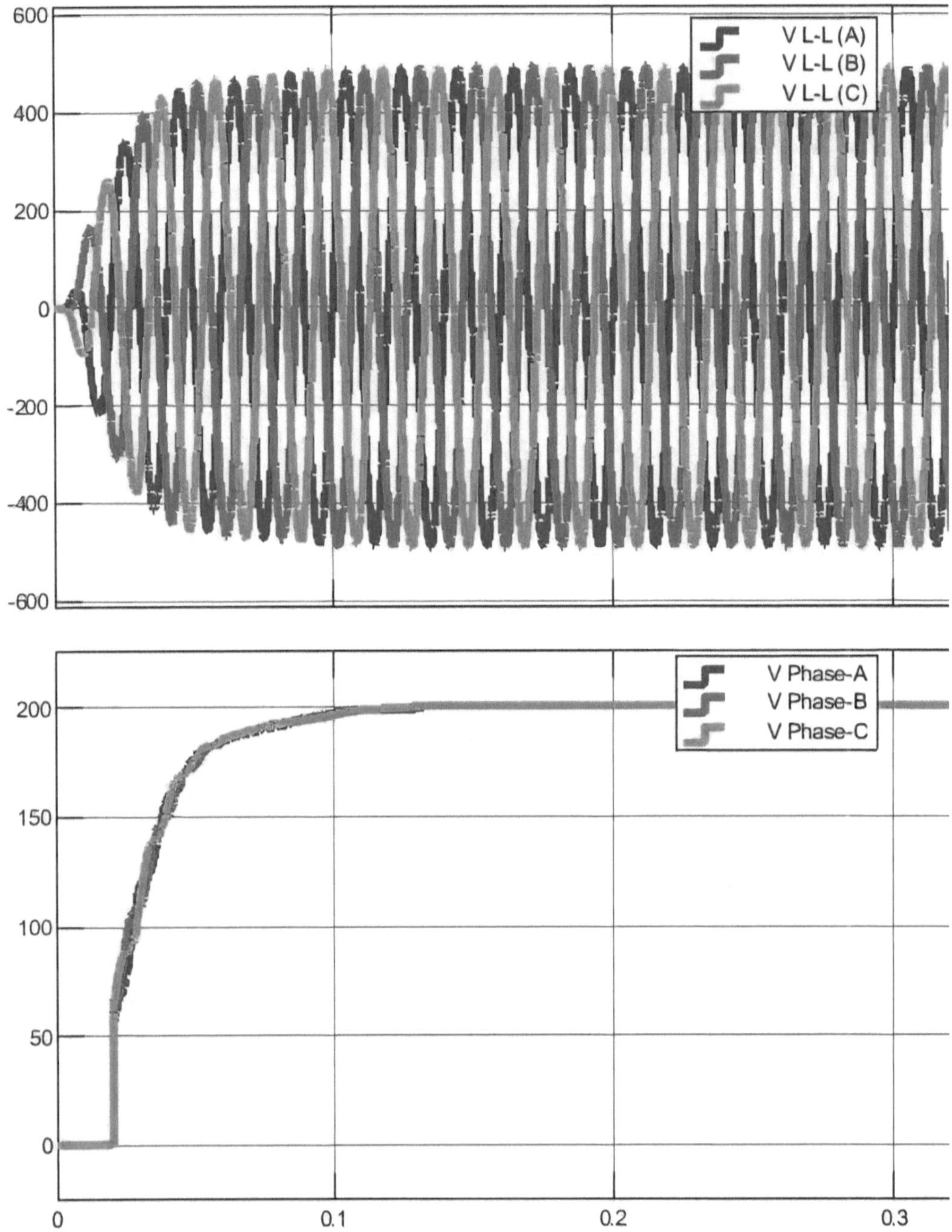

FIGURE 9.3 Inverter phase and line voltage output. (a) without DC Imbalancing (Kp= 20.6, Ki= 15) (continued)

9.3.3.2 Battery Storage System Integrated with a DC-DC Converter

The integration of a battery storage system with a DC-DC converter represents a dynamic and robust solution for efficiently managing and utilizing electrical energy. This system enables the bidirectional power flow between the battery and the DC bus, allowing for the

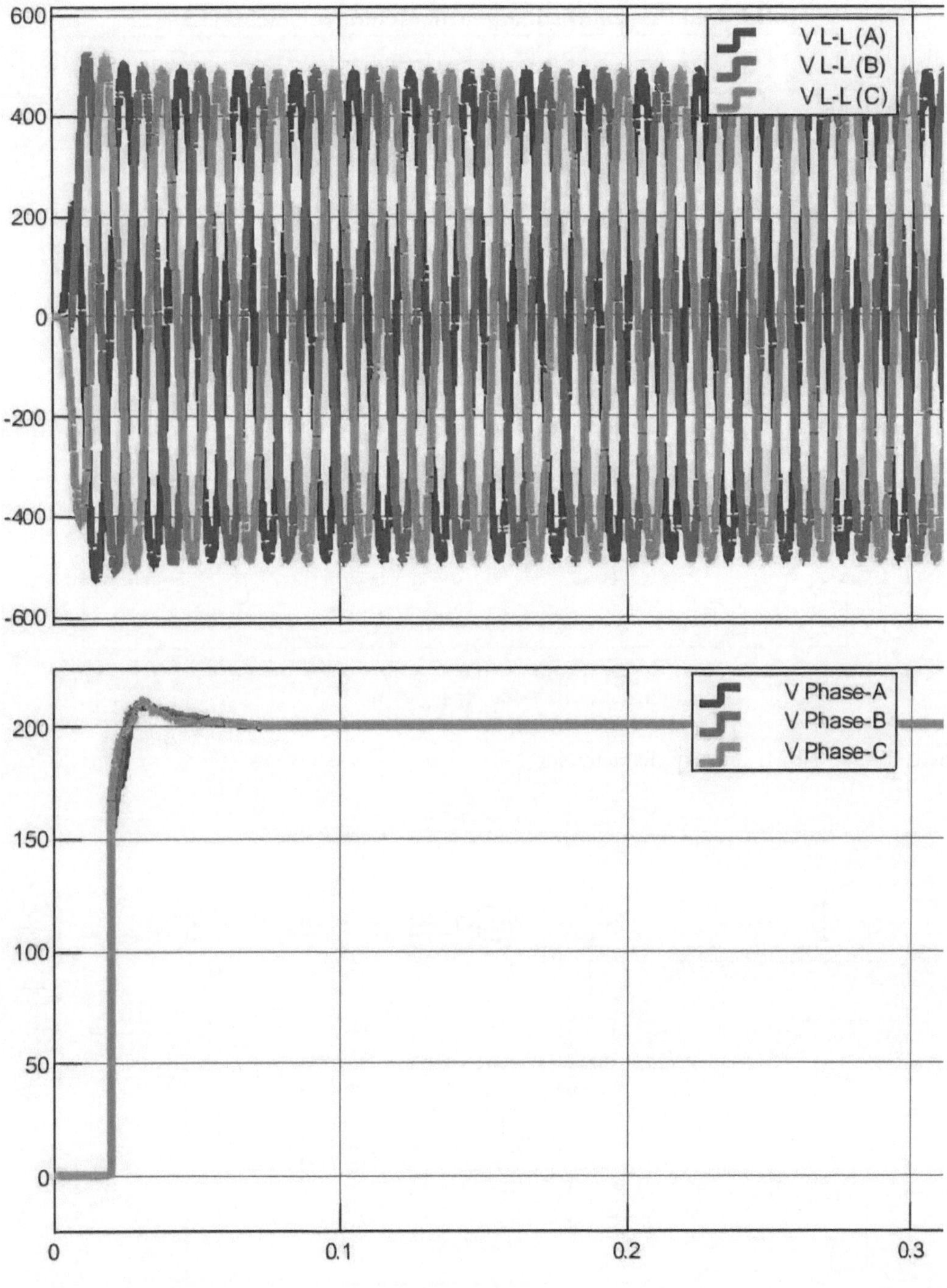

FIGURE 9.3 (Continued) Inverter phase and line voltage output. (b) DC Imbalancing (kp= 0, ki= 36).

storage of surplus energy during periods of excess generation, such as when renewable sources like solar panels or wind turbines produce more power than needed. Conversely, during increased demand or energy shortages, the stored energy can be discharged from the battery, and the DC-DC converter facilitates the conversion of the DC voltage levels

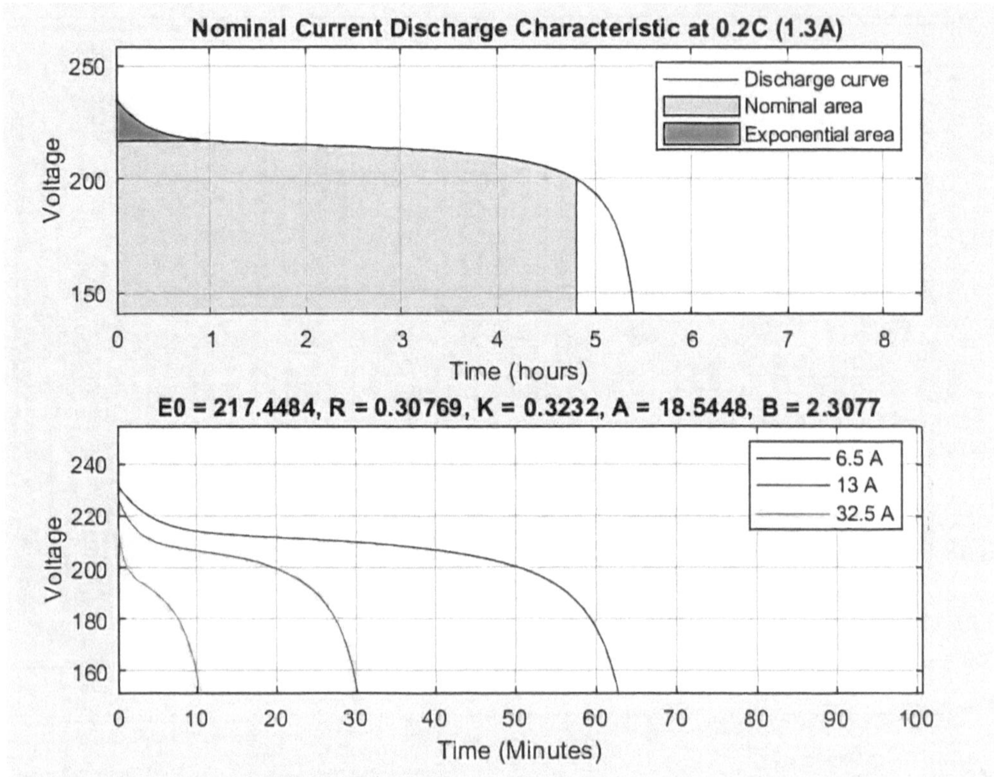

FIGURE 9.4 Ni-MH battery characteristic.

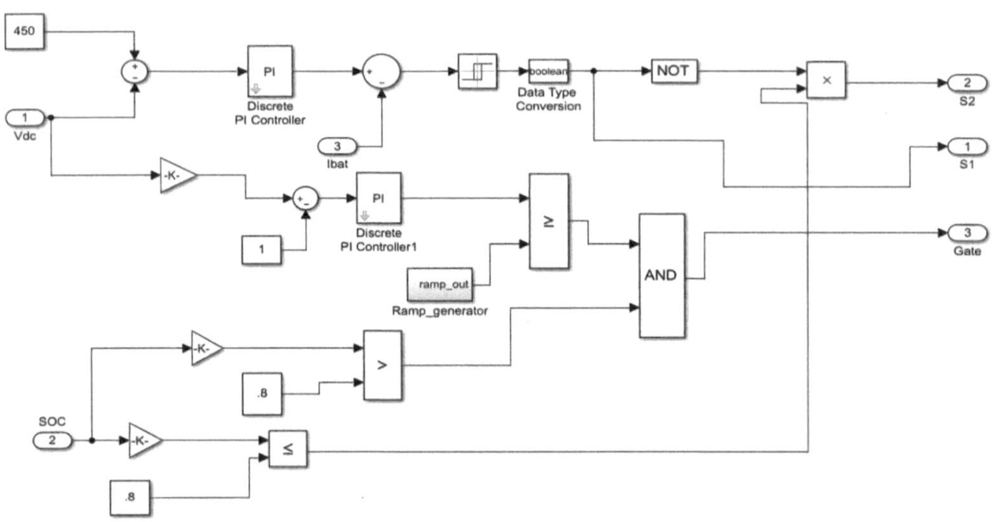

FIGURE 9.5 Battery controller for stable charging.

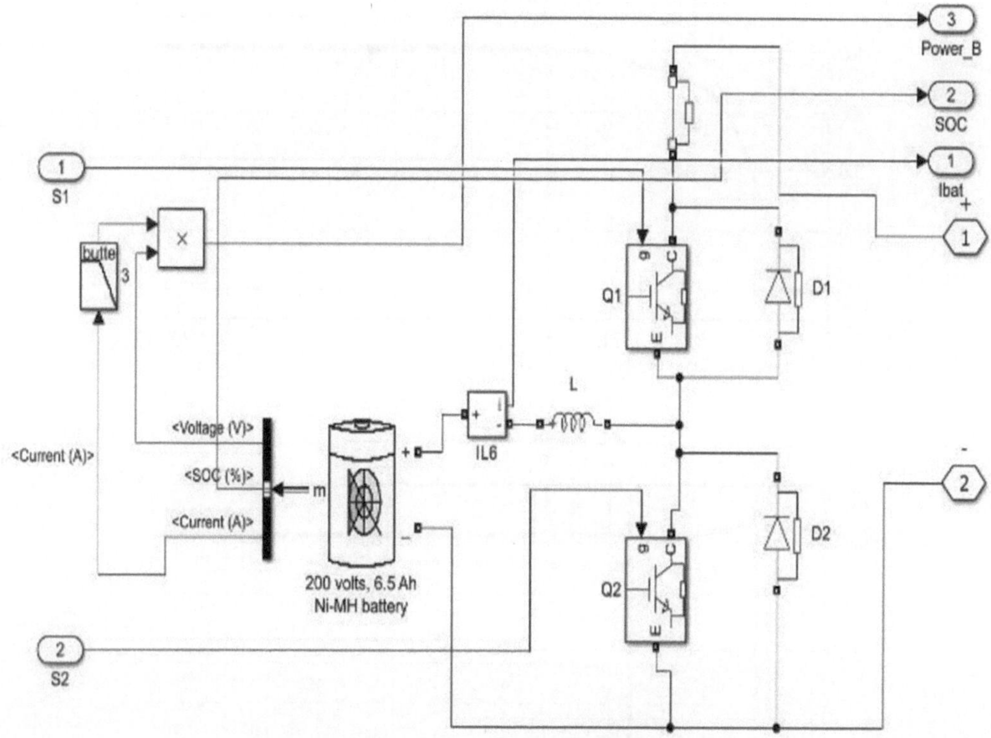

FIGURE 9.6 Battery storage system integrated with DC-DC converter.

to meet specific load requirements. This converter plays a pivotal role in maintaining the system's stability by regulating voltage and current levels, ensuring smooth energy transfer, and minimizing losses in the process. This integrated approach gives grid operators, businesses, and homeowners the flexibility to optimize energy utilization, enhance grid resilience, and reduce energy costs, making it an increasingly popular solution in the transition towards more sustainable and reliable energy systems. Battery storage system integrated with DC-DC converter uses IGBT technology to provide fast response and low switching losses, resulting in stable DC output current (I_bat), SOC of battery, and stable DC output voltage (Vdc) to the input terminal of inverter and also feedback to the battery controller as shown in Figure 9.6.

9.4 SIMULATION RESULTS

The proposed approach consists of a PI controller, which is used for the controlling mechanism. Both inverter and battery controllers use a PI controller to stabilize the output power and battery charging to energize the load efficiently without fluctuations, which may lead to severe consequences. This section illustrates the hybrid power produced with and without DC-link imbalance, and a comparison between these two is provided for better understanding.

Contrarily, this DC-link imbalance will result in fluctuating the overall output of the hybrid system, which will be applied to the inverter input to provide power to the load.

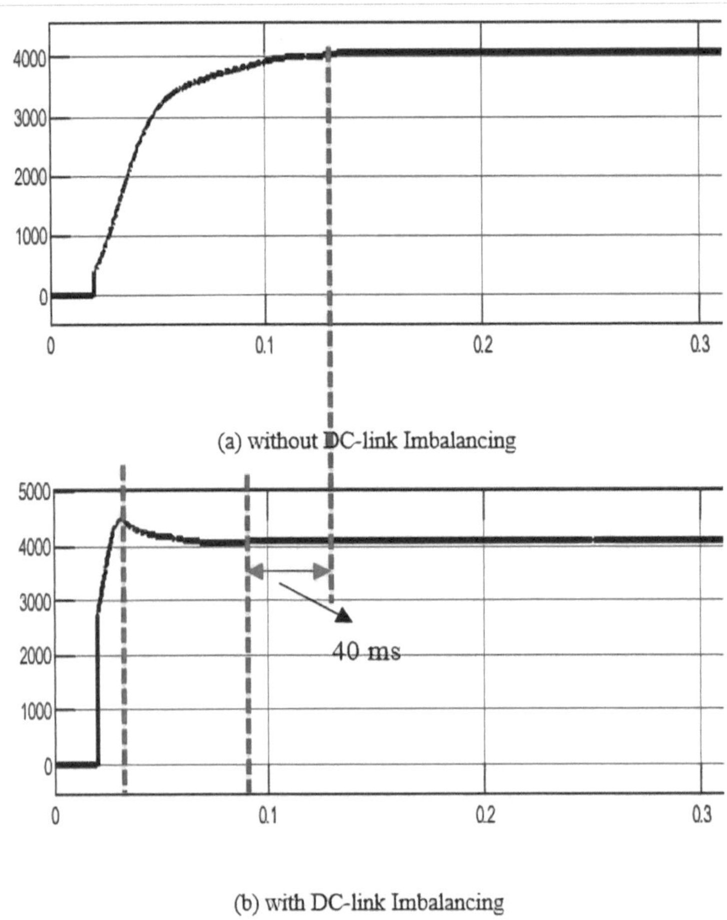

FIGURE 9.7 Hybrid power output.

The power output of the proposed 6 kW hybrid model simulates the wind output and solar output. For a better view, the overall hybrid output power (without DC-link imbalance) is illustrated in Figure 9.7(a).

However, when a Dc-link imbalance is applied to the model, we experience a quite sharp transition for the hybrid output at 50 ms, which takes approximately 30 ms to stabilize. Moreover, this sharp transition results from the wind power output fluctuations (which are observed during the experiment); after that, it is stabilized without any further fluctuations, as shown in Figure 9.7(b). This sharp transition can be further smoothened to provide a better waveform of output power to be fed to load. Note that our proposed model not only provides a stabilized hybrid power of 4.1 kW (without fluctuations) but also makes it more time-efficient, reducing the sharp transitions and making it smooth before 100 ms compared to without DC-link imbalance (40 ms faster).

Machine learning algorithms are implemented on training and testing datasets, respectively. Binary classification is performed on the overall simulation of XG Boost, Naïve Bayes, Complement Naïve Bayes, ADA Boost, and Gaussian NB. Figure 9.3 shows the normal and abnormal behavior of network traffic. Where around 75.99% is normal and 24.01% is considered abnormal.

These methodologies collectively aim to enhance grid-tied hybrid PV-Wind systems' operational efficiency, stability, and sustainability by leveraging AI techniques to tackle DC-link imbalances.

9.5 CONCLUSION & FUTURE DIRECTIONS

In this chapter, the hybrid power generation (i.e., wind and photovoltaic) model was studied and an optimized approach was presented to overcome the output fluctuation of the hybrid power generator to tie with the grid station. PI controller base for inverter and battery storage is presented. The concept of net metering also lies in sharing the extra power produced by the hybrid system with the grid to get some incentives. The proposed approach shows better results as compared to the existing schemes. The inverter controller stabilizes the hybrid power output and the inverter output, which is fed to the load. The comparison between the hybrid power output for both with and without DC-link imbalancing depicts that the proposed model efficiently stabilizes and provides the desired power output to the load in minimal time.

In future research, the stability of the hybrid system's power output can be further improved by applying AI algorithms. AI plays a central role in enhancing the performance of grid-tied hybrid PV-wind systems facing DC-link imbalances. Within these systems, AI acts as a sophisticated controller, maintaining constant vigilance over operations to uphold grid stability and maximize the utilization of renewable energy resources. By continuously analyzing real-time data, AI swiftly identifies imbalances in the DC-link resulting from the variable nature of solar and wind inputs, enabling rapid corrective measures to ensure a seamless power flow into the grid. Moreover, AI contributes to more accurate energy forecasting, facilitating the seamless integration of fluctuating renewable sources.

Additionally, AI optimizes various aspects of the system, including power conversion, load distribution, and interactions with the grid, all of which collectively enhance efficiency and reduce strain on system components. AI bolsters the system's resilience through its adaptive control capabilities and ability to detect faults. It ensures consistent and efficient energy generation while supporting the transition toward sustainable and grid-integrated hybrid energy solutions.

REFERENCES

[1] Shafiabady, N., Hadjinicolaou, N., Din, F. U., Bhandari, B., Wu, R. M., & Vakilian, J. (2023). Using artificial intelligence (AI) to predict organizational agility. *Plos one*, 18(5), e0283066.

[2] Ramli, M. A. M., Bouchekara, H. R. E. H., & Alghamdi, A. S. (2018). Optimal sizing of PV/wind/diesel hybrid microgrid system using multi-objective self-adaptive differential evolution algorithm. *Renewable Energy*, 121, 400–411.

[3] Sanjari, M. J., Gooi, H. B., & Nair, N. -K. C. (2020). Power generation forecast of hybrid PV–Wind system. *IEEE Transactions on Sustainable Energy*, 11(2), 703–712.

[4] Tak, S.Z., & Garg, V.K. (2018). Simulation of grid supported PV/wind/battery hybrid system with variable load. *2018 Fifth International Conference on Parallel, Distributed and Grid Computing (PDGC)*, 627–632.

[5] Anoune, K., Bouya, M., Astito, A., & Abdellah, A. B. (2018). Sizing methods and optimization techniques for PV-wind based hybrid renewable energy system: A review. *Renewable and Sustainable Energy Reviews*, 93, 652–673.

[6] Sba, M., Khedidja, B., Bakelli, Y., Kaabeche, A., & Khenfous, S. (2018). Sizing of a hybrid (Photovoltaic/Wind) pumping system based on metaheuristic optimization methods. In *Proceedings of the International Conference on Wind Energy and Applications* (pp. 1–6).

[7] Chen, Y.-M., Cheng, C.-S., & Wu, H.-C. (2006). Grid-connected hybrid PV/wind power generation system with improved DC bus voltage regulation strategy. In *Proceedings of the Twenty-First Annual IEEE Applied Power Electronics Conference and Exposition, 2006 (APEC '06)* (p. 7). Dallas, TX, USA.

[8] Jia, H., Xiao, Q., & He, J. (2018). An improved grid current and DC capacitor voltage balancing method for three-terminal hybrid AC/DC microgrid. *IEEE Transactions on Smart Grid*, pp. 1–1.

[9] Smith, J. (2021). Optimizing the DC-link imbalance in the grid-tied hybrid PV-wind system for artificial intelligence perspective. *Renewable Energy Journal*, 15(3), 127–143.

[10] Mane, S., Kadam, P., Lahoti, G., Kazi, F., & Singh, N. M. (2016). Optimal load balancing strategy for hybrid energy management system in DC microgrid with PV, fuel cell, and battery storage. In *Proceedings of the 2016 IEEE International Conference on Renewable Energy Research and Applications (ICRERA)* (pp. 851–856). Birmingham, UK.

[11] Brown, A. C. (2019). AI Applications in Grid Integration of Renewable Energy Sources. *IEEE Transactions on Sustainable Energy*, 7(2), 78–92.

[12] Garcia, L. M., & Patel, S. K. (2018). Artificial Intelligence Techniques for Improving Renewable Energy Grid Integration. In *Proceedings of the International Conference on Sustainable Energy Systems (ICSES)* (pp. 235–248). IEEE.

[13] Wang, L., Vo, Q. S., & Prokhorov, A. V. (2018). Stability improvement of a multimachine power system connected with a large-scale hybrid wind-photovoltaic farm using a supercapacitor. *IEEE Transactions on Industry Applications*, 54(1), 50–60.

[14] Bacchav, L., & Gaikwad, A. (2017). MATLAB implementation of standalone hybrid wind-solar power generation with and without dump power control. In *Proceedings of the International Conference on Electronics, Communication, and Data Science* (pp. 3069–3073).

[15] Qadir, Z., Khan, S. I., Khalaji, E., Munawar, H. S., Al-Turjman, F., Mahmud, M. P., ... & Le, K. (2021). Predicting the energy output of hybrid PV–wind renewable energy system using feature selection technique for smart grids. *Energy Reports*, 7(1), 8465–8475.

[16] Masood, B., Usman, M., Din, F.U. et al. Effect of transient and non-transient models on the performance of PLC. *Telecommunication Systems*, 65(1), 55–64 (2017). https://doi.org/10.1007/s11235-016-0211-1

[17] Hu, E., Wang, Z., Zhao, H., Guo, J., & Yang, H. (2017). A novel control strategy to smooth power fluctuation of hybrid offshore wind and tidal power generation system. *2017 IEEE Conference on Energy Internet and Energy System Integration (EI2)*, 1–4.

[18] Al-Turjman, F., Qadir, Z., Abujubbeh, M., & Batunlu, C. (2020). Feasibility analysis of solar photovoltaic-wind hybrid energy system for household applications. *Computers & Electrical Engineering*, 86(1), 106743.

[19] Zhang, H., et al. (2019). An AI-based predictive maintenance system for wind turbines in hybrid PV-wind power plants. *Renewable Energy*, 138(1), 758–769.

[20] Saeed, M. A., et al. (2018). Enhanced MPPT algorithm for a grid-connected hybrid PV-wind energy system using neural networks. *IEEE Access*, 6(1), 25048–25058.

[21] Wang, Y., et al. (2017). Artificial neural network-based fault diagnosis of hybrid wind-PV energy systems. *Energies*, 10(11), 1853.

[22] Abujubbeh, M., Marazanye, V. T., Qadir, Z., Fahrioglu, M., & Batunlu, C. (2019, June). Techno-economic feasibility analysis of grid-tied PV-wind hybrid system to meet a typical household demand: Case study-amman, Jordan. In *2019 1st Global Power, Energy and Communication Conference* (GPECOM) (pp. 418–423). IEEE.

[23] Islam, S., et al. (2015). Control and power management of a grid-connected hybrid wind/photovoltaic/fuel cell distributed generation system. *IEEE Transactions on Sustainable Energy*, 6(3), 697–708.

[24] Rasool, M. H., Perwez, U., Qadir, Z., & Ali, S. M. H. (2022). Scenario-based techno-reliability optimization of an off-grid hybrid renewable energy system: A multi-city study framework. *Sustainable Energy Technologies and Assessments*, 53(1), 102411.

[25] Huang, X., Sun, K., & Xiao, X. (2020). A neural network-based power control method for direct-drive wave energy converters in irregular waves. *IEEE Transactions on Sustainable Energy*, 11(4), pp. 2962–2971..

[26] Mujeeb Khan, N., Mansoor, M., Kumayl Raza Moosavi, S., Khan, K., Qadir, Z., & Hamza Zafar, M. (2022). Machine Learning-Based Maximum Power Point Tracking Technique for Concentrated PV/TEG System Under Non-uniform Environmental Conditions. In *Artificial Intelligence of Things for Smart Green Energy Management* (pp. 81–101). Cham: Springer International Publishing.

[27] Musunuri, S. K. (2011). *Control and interface design for cost reduction of a low power grid-connected wind-photovoltaic system*. Theses and Dissertations. 1175. https://scholarsjunction.msstate.edu/td/1175.

CHAPTER 10

IoT-Based Electrical Appliances

Basheer Ahmed, Zaigham Abbas, Abdullah Akbar, Usama Hussain Ahmad, and Shahab Ur Rehman

10.1 INTRODUCTION

These days, technology permeates every aspect of peoples' lives. The Internet is today commonly used for home automation, providing a range of services related to energy efficiency, safety, and security. Even at long distances, the user can watch over and control their house gate, different appliances, and turn the television on or off without being present. In spite of these benefits, home automation has yet to garner a lot of recognition and focus due to its great importance and intricacy [1-2]. an Internet of Things (IoT) network comprises interconnected sensors, computers, and digital devices that can communicate with one another to exchange and transport data. Each device has a UID (Unique Identifier). An IoT control system connected to a cloud server is utilized for early issue resolution and enhanced energy conservation [3].

The following are some common IoT applications:

1) A smart exam that uses IoT to gauge how difficult and disabled a student is to answer exam questions [4]. This technique makes it possible to assess a student's comprehension of a particular topic or subject.
2) An IoT-based remote patient health monitoring program [5].
3) Modern vehicle monitoring systems designed to track a vehicle's performance and get real-time input on its movement [6].

Due to the transfer of data in an open cloud infrastructure, IoT implementation requires careful consideration and diligence [7]. The act of managing or running different

DOI: 10.1201/9781032648323-10

types of machinery & other automation refers to appliances that use a variety of control methods and require little or no human intervention. Depending on the application, there are multiple categories of automation. IoT-based automation enables data sharing across connected devices and remote control of such devices. IoT uses Internet protocols or cloud-based computing to manage devices in places like smart homes. A "smart home" is an automated system that provides comfort, convenience, and energy efficiency. It is equipped with devices, a communication network, and sensors that can be automatically and remotely operated from any place using voice commands, a GSM network, or smartphone apps [8–12].

In addition, IoT is employed in sectors like healthcare, energy systems, smart automobiles, and industrial security.

IoT networks expand beyond physical objects in international telecommunications, encompassing virtual entities like content, creating an environment where objects and entities surround individuals, unlike M2M. M2M, in contrast, is focused on a single machine that acts as the main device, but IoT introduces an environment where people are surrounded by devices. The automation system is becoming more and more popular due to its many benefits. Just connecting electrical devices to the internet or cloud storage can create an automated system. The popularity of network-enabled automation systems has grown recently due to their low cost and ease of use. Cloud computing platforms facilitate connectivity between people and things, making it simple to access anything at anytime and anywhere through user-friendly interfaces that are specifically designed for that purpose. Thus, to access IoT, the cloud serves as a front end. By connecting and interacting, the IoT seeks to enhance the quality of human existence with any device or thing through the Internet without the need for human involvement, according to Ref. [13].

Smart homes use IoT technology. In earlier home automation systems the connection between an actuator and a sensor was clear [14], and the network structure included servers or gateways for information collecting and analysis as well as sensor nodes for data transmission and collection [15]. A smart home, often called an home or a linked home, is a living space with extremely sophisticated automated systems. A computer monitors a smart home's daily operations, giving the impression that it is "intelligent." A smart home uses a variety of technology connected to a home network to enhance living conditions. A smart house is equipped with sophisticated automated systems for managing and keeping an eye on many media devices, security systems, lighting and temperature control, and other features. IoT [16] is crucial to the construction of smart homes. Nearly every item we use on a daily basis in our homes may be connected to the Internet thanks to IoT. IoT makes it possible to keep an eye on and manage every one of these linked devices from anywhere at any time [17–23].

10.2 HARDWARE SECTION

PLCC, Esp8266 node MCU, Arduino Atmega328 all of these are hardware components. Therefore all of these comes under category of hardware.

10.2.1 PLCC Overview

Transfer line protection and voice data transfer are two applications for power line carrier communication, or PLCC. The frequency range of the carrier current utilized for power line carrier communication is 80–500 kHz. In modern residences and structures, transmitting digital information, especially audio, throughout the house and building is essential. However, the installation of new cables to enable this can pose challenges, making maintenance more complex and costly. PLCC is the best option to include in the automation concept in order to get around this, the shown in Figure 10.1. Automation systems, often referred to as smart technology, are a group of equipment and systems in a house or building that may communicate with one another or work independently to be as efficient as possible. Electrical appliances and the internet are connected via the home or building's current power wirings using PLCC technology. Using the current power cable infrastructure for communication is the fundamental concept behind PLCC. A variety of devices use PLCC technology. The primary advantage of this system is that with only a basic transmitter and receiver configuration and an equal phase supplying houses and buildings can be automated.

10.2.2 Esp8266 Node MCU Overview

The open source firmware and development kit Node micocontroller unit (MCU) is useful for creating or prototyping IoT products. It comes with software for Espressif Systems' ESP8266 Wi-Fi system on a chip (SoC) and hardware, which uses the ESP-12 module as a basis. An automation system describes a home or structure's capacity to make decisions on its own based on environmental factors. These systems can also be controlled remotely. To set up a web server using a NodeMCU or a similar ESP8266-based board for controlling GPIOs over Wi-Fi, begin by connecting the board to your Wi-Fi network and ensuring proper power supply. Develop firmware for the board with a built-in web server to handle HTTP requests and manage GPIO pins accordingly. Implement home or building automation logic within the firmware to control devices based on

FIGURE 10.1 PLCC.

FIGURE 10.2 Esp8266 Node MCU.

received commands. Concurrently, develop an Android application to send commands to the ESP8266 via Wi-Fi, using a defined communication protocol. Test the system thoroughly before deployment, ensuring seamless interaction between the Android app and the ESP8266 board for efficient home or building automation. The devices linked to home or building automation, as well as the Nodemcu esp32's built-in wifi module, are shown in Figure 10.2. An authentication token is used to connect both WiFi networks.

10.2.3 Arduino ATmega328 Overview

An 8-bit ATmega328P microprocessor serves as the foundation for the Arduino Uno microcontroller board. In addition to ATmega328P, it also includes other parts including a voltage regulator, a crystal oscillator, serial communication, etc., in order to help the microcontroller. A computer can connect with an Arduino device, alternative microcontrollers, or another Arduino board. Digital pins 0 and 1 can be used for UART TTL (5V) serial communication with the ATmega328P microcontroller (Tx), for example. The board's ATmega16U2 routes this serial connection over USB, appearing to computer software as a virtual com port. An open-source electronics platform called Arduino is built on user-friendly hardware and software. Arduino is an open source electronics single-board microcontroller and microcontroller kit (shown in Figure 10.3) that may be used to create interactive objects and digital and physical devices that can sense and control objects.

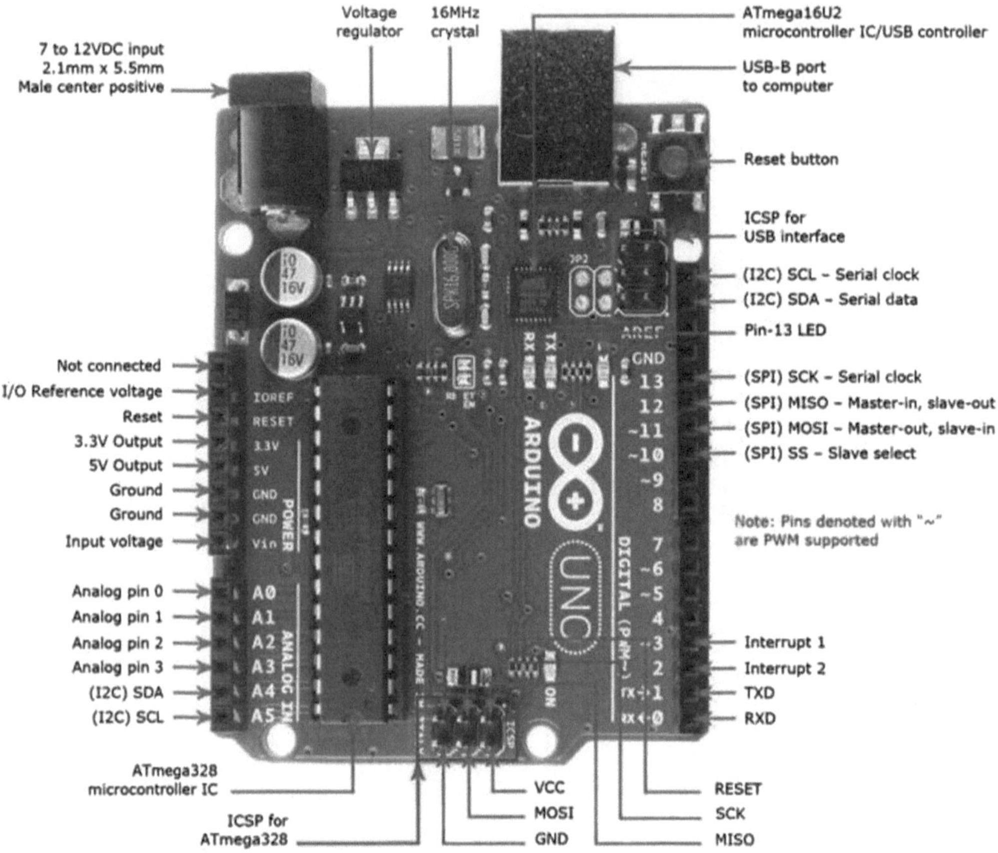

FIGURE 10.3 Arduino ATmega328.

The Arduino board is made with a range of controllers and microprocessors. The processing language project serves as the basis for the Arduino project's integrated development environment (IDE), which it uses in addition to conventional compiler toolchains.

10.2.4 Channel Relay Board

Computer boards featuring a variety of switches and relays are called relay boards. They are used to regulate the voltage supply and have input and output terminals. Relay boards offer several onboard relay channels that are individually programmable, offering real-time control. For remote device switching, a separate hardware component called a relay module is used. Using it, through the internet or a network, we may remotely operate devices. The AC adapter powers the module. Electronic device switching mechanisms are powered by relay boards. Circuitry for power supply and regulation is integrated into the system. Relay boards find applications across various products, serving to control lighting, motors, and other technological devices commonly used in commercial and industrial environments (refer to Figure 10.4).

FIGURE 10.4 Channel Relay Board.

10.2.5 Adapter

An adapter is a device that changes the characteristics of one system or device to another that would not otherwise work together. Some alter signal or power characteristics. Others only modify the external shape of one connection to another.

10.2.6 Regulator

A voltage regulator's job is to maintain a circuit's voltage somewhere near to the intended level. One of the most prevalent types of electrical components is the voltage regulator. A voltage regulator is a tool used to keep a power source's voltage within reasonable bounds. Regardless of changes in the input voltage or load state, it produces a set output voltage that stays constant. It serves as a buffer to keep device components safe.

10.3 RESULTS AND DISCUSSION

We created an IoT-based electrical appliance control system that can operate building and residential appliances in response to user commands. Additionally, it is intelligent enough to operate every device in accordance with the modes the user selects. It has the ability to allow the user to turn on and off particular appliances. This system's mode-based functioning is its most important feature. For example, all of the lights and the fan will intelligently switch on or off when the user activates auto mode.

10.3.1 Software Development

The project's software and hardware are interfaced using Arduino software. For the microcontroller to function and carry out the steps involved in the suggested design, it needs a program. After writing the code, verification and compilation are simple. Figure 10.6 shows the whole flowchart that details how the system operates. The comprehensive flowchart demonstrates how the system operates as a whole and is managed using a mobile application. The goal of this project is to automate buildings or homes. First, connecting to the current network infrastructure, the Wi-Fi shield (Wi-Fi hotspot) initializes the open-source server Blynk. When a program that serves the client (operator) receives a signal from the Wi-Fi module indicating that the system is online or offline, it checks the input-output pins, or switch 1, 2, and 3. The data is received by the Blynk server if the client (operator) flips any switches, and it displays the status to the user on the in-app LCD. Finally, the system cycles back to the starting state while this procedure runs continuously.

10.4 SUMMARY OF NODE MCU CODING

There are open source prototype board designs available for the NodeMCU firmware. Users can easily access a business's Wi-Fi connectivity from the login or admin area by selecting it from the list of available connections. However, to establish a connection, users are required to input security codes or passwords. The list of connections includes options such as "Business_WiFi," "Company_Wireless," "Office_Network," "Corporate_WiFi,"and "Enterprise_Network". The "NodeMCU" refers to the combination of the "Node" and "MCU" (microcontroller unit). Strictly speaking, "NodeMCU" refers to the firmware rather than the related development kits.

Wi-Fi is a radio-based wireless network. WLAN (Wireless LAN), WiFi, WiFi, "wireless network," or plain "wireless" are common names for Wi-Fi. An access point is required for a wireless LAN so that PCs, phones, and other devices are able to connect. In a typical home network, usually the router serves as the access point. A business's Wi-Fi connectivity may be easily accessed by the user(s) in the (Login/Admin area) by choosing it from their list of connections. Security codes or passwords are necessary.

- Step 1: Include file header which means the library of any software code.
- Step 2: Give datatype of my program Datatype is int, char, string, etc
- Step 3:in step three made the program of Admin/Login Username and password.
- Step 4: The application launches the Server main function once the user inputs their name and password.
- Step 5: If the user inputs a valid username and password then the compiler executes the program and gives the output but if a user enters an invalid username or password then compile Runs but show the message of invalid input.
- Step 6: Using a while loop to connect to a WiFi server runs continuously until the condition becomes false.
- Step 7: After Connect the user, the Output message will be shown on the mobile. application screen.
- Step 8: The output will be connected to Wi-Fi valid inputs and IP address.
- Step 9: Condition for If any client is connected then show output on the screen.

10.5 SUMMARY OF ARDUINO CODING

AT_MEGA_328p:-

AT_Mega_328p is Microchip's high-performance, low-power controller.

AT_MEGA_328p is an 8-bit microcontroller using AVR RISC architecture as its foundation. Because it is incorporated into Arduino boards, it is the most widely used AVR controller.

Code of Arduino AT mega 328p

One by One Step Of Arduino AT mega 328p:-

- Step 1: Include Libraries Such As:- (#include<SoftwareSerial.h>)
- Step 2: Develop a function with parameters to control the LED lights. When the condition evaluates to TRUE, the function turns all the LEDs on, representing the HIGH state. Conversely, when the condition becomes FALSE, the function turns all the LEDs off, indicating the LOW state. Finally, ensure that the results of these actions are displayed on the output screen.
- Step 3: Made Light Emitting Diode called (LED) having (4) Relay/s into a String Datatype.
- Step 4: Create Another function called (SETUP) with Void/(Empty) which means not a Return value.
- Step 5: Call the Function in (Main) Reserved Word/s Like:- Serial.begin(9600); PLComm.begin(9600); pinMode(led,OUTPUT);
- Step 6: Calling All the relay/s with the delay of (1000) Basically Delay is a Reserved word Which means (built_in_function)Delay Means "A time by which something late or postponed."
- Step 7: Output/Show Message on the screen "Welcome" ("\n" For new Line.)
- Step 8: Using the While loop and with the delay of(5). Loop executes until Condition is false.
- Step 9: Show a message on the Output Screen. Create another Condition. If(msg.length()>50) The digital relay light begins Low and High And Low Means "OFF" And High means "ON"
- Step 10: When the condition is TRUE, all LED lights switch to the ON state, indicating HIGH, and when the condition is FALSE, they switch to the OFF state, indicating LOW. The results of these actions are displayed on the output screen..

10.6 MOBILE APPLICATION

We created the Android application Smart Control for our system. With Smart Control, users can manage every aspect of the system. A user's can also set modes. Building automation and smart homes are in high demand, whereby appliances are readily managed by a single, shared device and respond automatically to changing environmental circumstances. The automated system saves time and energy and contributes to a reduction in the need for human labor.

The app interface is shown in Figure 10.5 and Figure 10.6. The login page, which uses the user's IP address to authenticate, is located on the app's home screen; following a successful login the user's mobile app will allow them to operate any electrical appliance in the home. We used the MIT App Inventor to create this application. Mit App Inventor

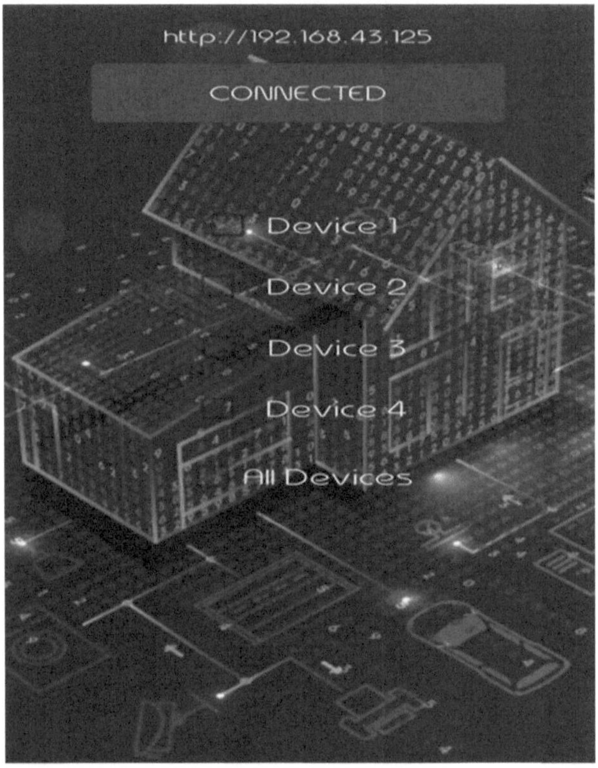

FIGURE 10.5 User interface of the smart control application.

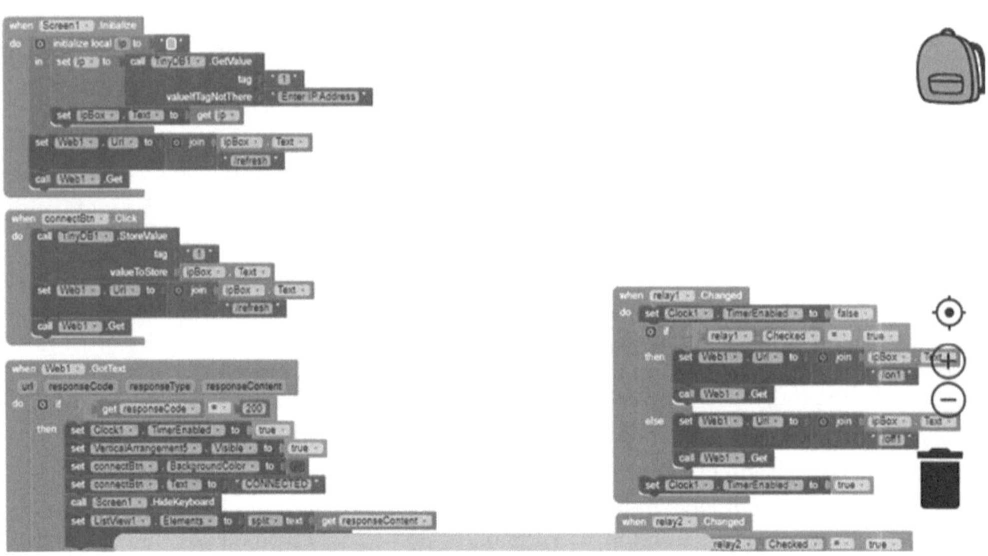

FIGURE 10.6 Blocks diagram of app.

is a free platform that allows programmers to create practical Android apps. Block coding is used in MIT App Inventor to construct the back-end functions. It is simple to create algorithms for block codes. Smart Control block codes are made such that, when pressing a button, they will connect to the IP address the user enters. That IP address is the network-connected IP address of the NodeMCU. This is how Smart Control's back-end functions are set up.

10.7 IMPLEMENTATION OF THE SYSTEM

Throughout the implementation process, the automation system's transmitter and receiver were tested. We utilized our node MCU, PLCC as input, and Arduino AT mega328p as controller for the automation system, and lights as appliances. We verified whether or not the requirements for turning on lights were met, as seen in Figure 10.7. We went on to the next phase, which is controlling over the internet, once we were successful in this one (Figure 10.8).

10.8 CONCLUSION

This chapter has shown how to connect electrical appliances to our automation system via the IoT and control them remotely with a high percentage of accuracy. Users just need to choose the desired modes on the app; the appliances will be controlled by our system automatically.

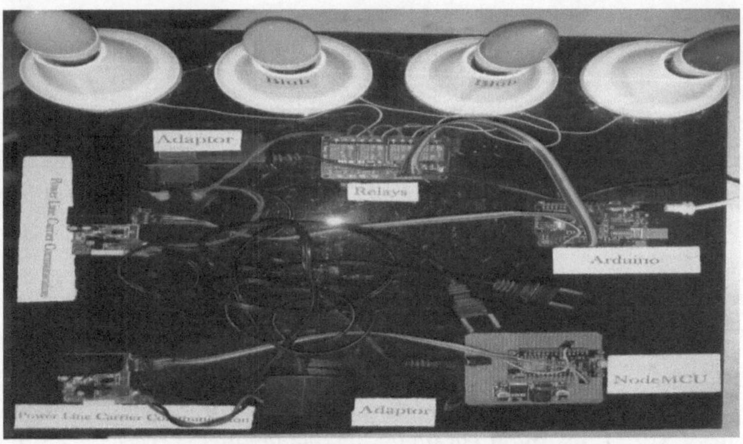

FIGURE 10.7 Implementation of automation system.

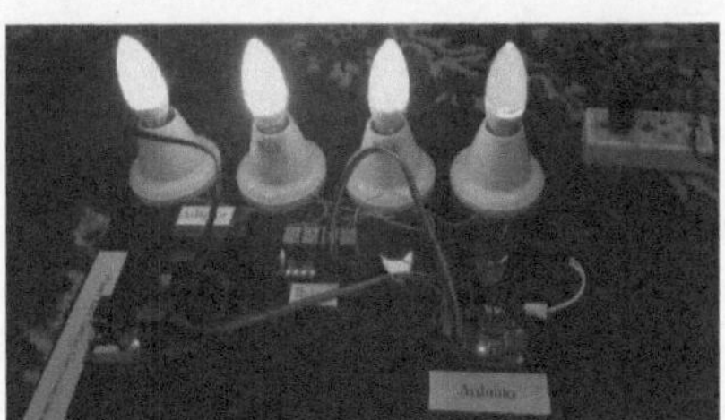

FIGURE 10.8 Devices are controlled by a smartphone with a smart control app.

10.9 FUTURE WORK

There are several improvements that could be made to this system in order to increase the sensing and detection accuracy:

a) More IoT sensors could be used to improve the home's security and management, such as a pressure sensor placed outside to sense whether someone is going inside.

b) Modifying automatic alerts to make this system more professional by utilizing the GSM module.

c) A smart garage to direct the automobile through the garage to make parking easier for the homeowner by measuring the length of the vehicle.

REFERENCES

[1] P. A. Dhobi and N. Tevar, "IoT based home appliances control," *2017 International Conference on Computing Methodologies and Communication (ICCMC)*, 2017. doi:10.1109/iccmc.2017.8282546

[2] A. R. Al-Ali and M. AL-Rousan, "Java-based home automation system," *IEEE Transactions on Consumer Electronics*, vol. 50, no. 2, pp. 498–504, 2004. doi:10.1109/tce.2004.1309414

[3] A. K. Gupta and R. Johari, "IoT based electrical device surveillance and Control System," *2019 4th International Conference on Internet of Things: Smart Innovation and Usages (IoT-SIU)*, 2019. doi:10.1109/iot-siu.2019.8777342

[4] A. Xheladini, S. Deniz Saygili, and F. Dikbiyik, "An IoT-based smart exam application," *IEEE EUROCON 2017 -17th International Conference on Smart Technologies*, 2017. doi:10.1109/eurocon.2017.8011164

[5] D. Minoli, K. Sohraby, and B. Occhiogrosso, "IoT security (IoTSec) mechanisms for e-health and ambient assisted living applications," *2017 IEEE/ACM International Conference on Connected Health: Applications, Systems and Engineering Technologies (CHASE)*, 2017. doi:10.1109/chase.2017.53

[6] S. Wang, Y. Hou, F. Gao, and X. Ji, "A novel IoT access architecture for vehicle monitoring system," *2016 IEEE 3rd World Forum on Internet of Things (WF-IoT)*, 2016. doi:10.1109/wf-iot.2016.7845396

[7] A. R. Biswas and R. Giaffreda, "IoT and cloud convergence: Opportunities and challenges," *2014 IEEE World Forum on Internet of Things (WF-IoT)*, 2014. doi:10.1109/wf-iot.2014.6803194

[8] Mahindrakar, Smita, and Ravi K. Biradar "Internet of things: Smart home automation system using Raspberry Pi," *International Journal of Science and Research (IJSR)*, vol. 6, no. 1, pp. 901–905, 2017. doi:10.21275/art20164204

[9] N. A. Prasetyo, A. G. Prabawati, and S. Suyoto, "Smart home: Power electric monitoring and control in Indonesia," *International Journal of Interactive Mobile Technologies (iJIM)*, vol. 13, no. 03, p. 143, 2019. doi:10.3991/ijim.v13i03.10070

[10] J. P. Sipani, R. H. Patel, T. Upadhyahya, and A. Desai, "Wireless sensor network for monitoring & control of environmental factors using arduino," *International Journal*

of Interactive Mobile Technologies (iJIM), vol. 12, no. 2, p. 15, 2018. doi:10.3991/ijim.v12i2.7415

[11] T. Eric, S. Ivanovic, S. Milivojsa, M. Matic, and N. Smiljkovic, "Voice control for smart home automation: Evaluation of approaches and possible architectures," *2017 IEEE 7th International Conference on Consumer Electronics - Berlin (ICCE-Berlin)*, 2017. doi:10.1109/icce-berlin.2017.8210613

[12] I. S. Areni *et al.*, "IoT-based of automatic electrical appliance for smart home," *International Journal of Interactive Mobile Technologies (iJIM)*, vol. 14, no. 18, p. 204, 2020. doi:10.3991/ijim.v14i18.15649

[13] C. Perera, A. Zaslavsky, P. Christen, and D. Georgakopoulos, "Context aware computing for the internet of things: A survey," *IEEE Communications Surveys & Tutorials*, vol. 16, no. 1, pp. 414–454, 2014. doi:10.1109/surv.2013.042313.00197

[14] K. Chang Lee and H. -H. Lee, "Network-based fire-detection system via controller area network for smart home automation," *IEEE Transactions on Consumer Electronics*, vol. 50, no. 4, pp. 1093–1100, 2004. doi:10.1109/tce.2004.1362504

[15] D. Han and J. Lim, "Smart home energy management system using IEEE 802.15.4 and zigbee," *IEEE Transactions on Consumer Electronics*, vol. 56, no. 3, pp. 1403–1410, 2010. doi:10.1109/tce.2010.5606276

[16] J. Gubbi, R. Buyya, S. Marusic, and M. Palaniswami, "Internet of things (IoT): A Vision, architectural elements, and future directions," *Future Generation Computer Systems*, vol. 29, no. 7, pp. 1645–1660, 2013. doi:10.1016/j.future.2013.01.010

[17] B. Davidovic and A. Labus, "A smart home system based on sensor technology," *Facta universitatis - series: Electronics and Energetics*, vol. 29, no. 3, pp. 451–460, 2016. doi:10.2298/fuee1603451d

[18] T. Malche and P. Maheshwary, "Internet of things (IoT) for building smart home system," *2017 International Conference on I-SMAC (IoT in Social, Mobile, Analytics and Cloud) (I-SMAC)*, 2017. doi:10.1109/i-smac.2017.8058258

[19] I. Ullah Khan, M. Ouaissa, M. Ouaissa, Z. Abou El Houda, and M. Fazal Ijaz, *Cyber Security for Next-Generation Computing Technologies*, 2023. doi:10.1201/9781003404361

[20] I. Ullah Khan, M. Abul Hassan, M. Fayaz, J. Gwak, and M. Adnan Aziz, "Improved sequencing heuristic DSDV protocol using nomadic mobility model for FANETS," *Computers, Materials & Continua*, vol. 70, no. 2, pp. 3653–3666, 2022. doi:10.32604/cmc.2022.020697.

[21] I. U. Khan, M. Y. Ayub, A. Abdollahi and A. Dutta, "A hybrid deep learning model-based intrusion detection system for emergency planning using IoT-network," 2023 *International Conference on Information and Communication Technologies for Disaster Management (ICT-DM), Cosenza, Italy, 2023*, pp. 1–5, doi: 10.1109/ICT-DM58371.2023.10286954.

[22] A. Abdollahi and M. Fathi, "An intrusion detection system on Ping of death attacks in IoT Networks," *Wireless Personal Communications*, vol. 112, no. 4, pp. 2057–2070, 2020. doi:10.1007/s11277-020-07139-y.

[23] I. U. Khan et al., "RSSI-controlled long-range communication in secured IoT-enabled unmanned aerial vehicles," *Mobile Information Systems*, vol. 2021, pp. 1–11, 2021. doi:10.1155/2021/5523553.

Index